Le Roy Clark Cooley

The New Text-Book of Physics

An elementary course in natural philosophy

Le Roy Clark Cooley

The New Text-Book of Physics
An elementary course in natural philosophy

ISBN/EAN: 9783337068974

Printed in Europe, USA, Canada, Australia, Japan

Cover: Foto ©berggeist007 / pixelio.de

More available books at **www.hansebooks.com**

THE NEW
TEXT-BOOK OF PHYSICS

AN

ELEMENTARY COURSE IN NATURAL PHILOSOPHY

DESIGNED FOR USE IN

HIGH SCHOOLS AND ACADEMIES

BY

LE ROY C. COOLEY, PH.D.
PROFESSOR OF PHYSICS AND CHEMISTRY IN VASSAR COLLEGE

IVISON, BLAKEMAN, TAYLOR, AND COMPANY,
NEW YORK AND CHICAGO.

COPYRIGHT, 1868.

COPYRIGHT, 1880,
BY LE ROY C. COOLEY.

PREFACE.

THE Text-book of Natural Philosophy, first published in 1868, is now thoroughly revised; and an essentially New Text-Book of Physics is offered with the hope that it will prove to be still more worthy of the esteem of the friends of sound learning, and still more useful to the very large number of science-teachers who have so long employed the older book.

That which chiefly distinguishes this New Text-Book is the prominence which it gives to the principle of energy. Physics has come to be universally regarded by scientists as "the science of matter and energy;" and I have felt, that, unless I could fairly present the principle of energy as the most vital element of the system, my work would be altogether out of harmony with the later views, and unworthy to be used in the education of the young, because it would fail to give the student any adequate idea of the present state of the science.

Various changes in the arrangement of subjects have been accordingly made; and the doctrine of energy, in an elementary form such as it must take in so elementary a work, will be found giving tone to all departments of the course. The first three chapters, treating of MATTER and MOTION, will abundantly prepare the student for the study of ENERGY in the fourth; while, in the remaining chapters, he will be able to trace the exhibition of energy as seen in the phenomena of sound, heat, light, electricity, and machinery.

Much new matter has been introduced, as, for example, on the subject of electrical induction, the telephone, the phonograph, and other discoveries and applications which the late progress of science has made fit subjects for an elementary course.

Among the new features designed to adapt the work to the actual wants of the class-room, is the Review at the end of every chapter. These reviews consist of Summaries of Principles, Summaries of Topics, and Problems.

The *summaries of principles* are not simply a re-statement of principles in the order of their previous discussion: they are constructed in a way to show these truths in new relations, and to supplement the synopses of the paragraphs in giving clear but concise statements of the most vital principles of the science. The *summaries of topics* follow, and suggest to the mind all the points in the discussion of the subject of every paragraph. And finally the problems tend to make the student's knowledge more exact by requiring him to apply it mathematically.

The *References* which accompany the synopsis of each subject are specific, not simply to a book by title, but to the particular paragraph of the book where the desired information may be found; and not to many and inaccessible authors, but, throughout, to two, one or the other of which, or both, may be in the hands of the teacher or pupil. Ganot's Physics is selected because of its excellent and thorough discussions, and Arnott's because of the appropriate and interesting illustrations in which it abounds. These references are designed for the *actual use* of the student who is advanced enough and has time to extend his study beyond the elementary course in this work, and of the teacher who desires to bring additional matter into the class-room to elucidate or extend these elementary texts.

If circumstances permit the use of a larger number of

reference-books, I would mention Deschanel's Natural Philosophy, by Everett; Lardner's volumes on Mechanics and Optics; Silliman's Physics; Olmstead's Natural Philosophy, by Kimball. Mayer's works on Sound and Light; Sylvanus P. Thompson's Electricity and Magnetism; and Tyndall's several volumes on Sound, Heat, Light, and Electricity, — as among those likely to prove most generally useful.

The character of the work is still further indicated by the following extracts from the preface to the first edition: —

This volume is designed to be a *text-book* of natural philosophy suited to the wants of high schools and academies.

The author believes that the following features of his work adapt it to the purpose for which it is designed: —

1. It contains no more than can be *mastered* by average classes in the time usually given to this science.

2. It presents a judicious selection of subjects. Omitting whatever is merely novel or amusing, it gives a plain and concise discussion of *elementary* principles, of theoretical and practical value.

3. It is an expression of modern theories. It recognizes the fact that the spirit of a new philosophy pervades every department of science, and presents the doctrines of molecules and of molecular motions, instead of the old theory of imponderables, which has been swept away. Carefully avoiding whatever is yet only probable, it seizes upon what has come to be universally accepted, and, as far as may be, adapts it to the course of elementary instruction which it proposes.

4. It is logical in the arrangement and development of subjects. A single chain of thought binds the different branches of the science into one system of related principles.

5. It is thoroughly systematized. Chapters, sections, paragraphs, and topics have been arranged with careful regard, on the one hand, to the relation of principles to

each other, and, on the other hand, to the best methods of conducting the exercises of the class-room.

At the beginning of each paragraph is a plain and concise statement of useful facts and principles, while the paragraph itself contains the discussion of them by topics in their natural order.

There is an increasing number of teachers who believe that oral instruction is quite as important to the pupil as the study of a text-book. These headings of the paragraphs are *texts*, which, taken together, give a compact view of the entire science, and which will enable the teacher to freely supplement the discussions of the book, by experimental or mathematical proofs. To facilitate this work still further, references have been given to the most accessible and reliable works wherein the subjects of the text are more exhaustively treated.

LeR. C. C.

June, 1880.

CONTENTS.

CHAPTER I.

ON MATTER AND FORCE.

Section.	Page.
I. — Natural Philosophy Defined	1
II. — On Matter and Force	6
III. — On the Fundamental Ideas	16
IV. — Review	17

CHAPTER II.

ON THE THREE PHYSICAL FORMS OF MATTER.

I. — On the Application of the Fundamental Ideas,	20
II. — On the Characteristic Properties of Solids	21
III. — On the Characteristic Properties of Liquids	24
IV. — On the Pressure of Liquids	26
V. — On the Characteristic Properties of Gases	43
VI. — On Atmospheric Pressure	48
VII. — On "The Three Laws" for Gases	57
VIII. — Review	62

CHAPTER III.

ON MOTION.

I. — On Motion Produced by a Single Force	67
II. — On Motion Produced by more than one Force	77
III. — On the Motion of Liquids	86
IV. — On the Motion of Air	90
V. — On Vibration	92
VI. — On Undulations	103
VII. — Review	109

CHAPTER IV.

ON ENERGY.

SECTION. PAGE.
I.— On Definitions and Measures 114
II.— On the Conservation of Energy 123
III.— On the Recognition of Energy by the Senses . 126
IV.— Review 129

CHAPTER V.

ON MOLECULAR ENERGY, OR HEAT.

I.— On Conduction and Convection . . . 132
II.— On the Effects of Heat 135
III.— Review 146

CHAPTER VI.

ON UNDULATORY ENERGY, OR SOUND.

I.— On Transmission and Reflection of Sound . . 148
II.— On Musical Sounds 153
III.— On Musical and Sensitive Flames . . . 163
IV.— Review 167

CHAPTER VII.

ON RADIANT ENERGY, OR LIGHT.

I.— On Transmission 169
II.— On Reflection 172
III.— On Refraction 184
IV.— On Dispersion 194
V.— On Optical Instruments 209
VI.— On Double Refraction and Polarization . . 216
VII.— Review 220

CHAPTER VIII.

ON ELECTRICAL ENERGY.

I.— On Frictional Electricity 224
II.— On Magnetic Electricity 241
III.— On Dynamic Electricity 247
IV.— Review 269

CHAPTER IX.

ON MACHINERY

SECTION. PAGE.
I. — ON THE SIMPLE MACHINES 274
II. — ON WATER-POWER 293
III. — ON THE STEAM-ENGINE 296
IV. — REVIEW 300

METRIC MEASURES.

NAMES AND VALUES OF SMALLER UNITS.

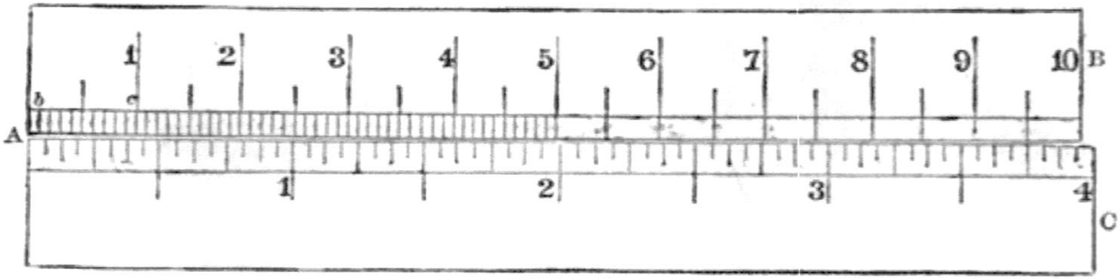

Fig. 1.

In Fig. 1, from A to b is 1 millimeter (mm.).
" " A " c " 10 " = 1 centimeter (cm.).
" " A " B " 100 " = 10 " = 1 decimeter = $\frac{1}{10}$ meter (m.).
" " A " C " 4 inches.

10 centimeters = 3.937 inches, or *almost* 4 inches.

∴ $\frac{\text{centimeters} \times 2}{5}$ = inches, nearly.

∴ $\frac{\text{inches} \times 5}{2}$ = centimeters, nearly.

1 cubic centimeter is the *volume* of a little cube measuring 1 cm. on a side.
1 gram = the weight of 1 cubic centimeter of pure water = 15.4 grains, nearly.

x

PHYSICS,

OR

NATURAL PHILOSOPHY.

CHAPTER I.

ON MATTER AND FORCE.

SECTION I.

NATURAL PHILOSOPHY DEFINED.

1. THE qualities of matter are usually called its properties. Extension, Impenetrability, Indestructibility, and Elasticity are examples.

The Properties of Matter. — In what respects is a block of granite so unlike a block of wood? The granite is brittle; it may be chipped with a chisel: the wood is soft; it may be cut with a knife. The granite is heavy; to lift it may require the power of an engine: the wood is much lighter; perhaps a single arm is able to move it. We are thus able to perceive a difference in bodies only because there is a difference in the qualities they possess. These *qualities* are called PROPERTIES.

Extension. — Every body of matter, however small, fills a portion of space. It is not possible to think of a body which should have no size. *This property of matter, by virtue of which it occupies space*, is called EXTENSION.

Measurement of Extension. — Every body of matter must have length, breadth, and thickness. The amount of space which a body fills is found by measuring these three dimensions. In England and the United States the *yard* is adopted as the unit of length. Feet, inches, rods, and miles are the divisions and multiples of the yard.

In France and many other countries the *metric measures* are employed, in which the unit is called the *meter*. The meter is the forty-millionth part of that meridian of our globe which passes through Paris. It is equal to 39.37 inches.

In the metric system smaller units are obtained by dividing the meter into tenths, hundredths, and thousandths; and larger ones, by multiplying it by ten, one hundred, and one thousand. The names and values of the smaller units are given on page x, Fig. 1.

Volume. — The amount of space which a body occupies is called its Volume. For bodies of small size the volume is measured in cubic inches, or, by the metric system, in cubic centimeters. For bodies of larger size a larger unit is more convenient; and their volumes are measured in cubic feet, or cubic yards, and by the metric system, in cubic meters.

Impenetrability. — Not only do all bodies occupy space: every body fills the space assigned it to the exclusion of all others. One body may not be pushed into the substance of another: it can take the place of another only when the other has been thrust away. When, for example, a nail is driven into wood, it pushes the particles of wood out of its way; and, when the hand is plunged into water, the water is thrust aside to give it place. *This property of matter, by virtue of which no two bodies can fill the same space at the same time*, is called Impenetrability.

Indestructibility. — A piece of gold may be cut into parts so small as to be almost invisible. It may be dissolved by acids, and made to disappear; or by intense heat it may be changed into thin vapor, and hid in the air. After all these changes have been wrought upon the gold, its particles may be again collected to form a mass like the original one without the slightest diminution in weight. Amid all the changes which we witness in the forms and qualities of bodies, not a single atom is destroyed. *This property of matter, by virtue of which no particle can be destroyed*, is called INDESTRUCTIBILITY.

Elasticity. — When an India-rubber ball is pressed in the hand, it is made smaller; but, the moment the pressure is removed, the ball springs back to its original size. The same quality is possessed in various degrees by all bodies. In such as lead and clay it is very slight, yet a ball made of either of these substances will spring back after having been for a moment compressed. On the other hand, an ivory ball, when let fall upon a marble slab, rebounds nearly to the height from which it fell, showing that the power of restitution is, in this case, almost equal to the force of compression. *This property of matter, by virtue of which it restores itself to its former condition after having yielded to some force*, is called ELASTICITY.

Are all Bodies elastic? — This property of matter is possessed by all bodies. Some are very slightly elastic; such are lead and clay. Glass, although very brittle, is highly elastic. A glass ball will rebound from a marble slab almost as well as one of ivory. Steel is likewise hard and brittle; yet the Damascus sword, which was made of steel, could be bent double without breaking.

But, should we attempt to describe all the properties of matter in detail, the time given to the study of our science would be filled with little else. The success of a student of nature depends largely upon his power to classify phenomena, and to study them in groups.

2. All the properties of matter may be grouped in two divisions, viz.: **Physical Properties**, of which malleability and ductility are examples; and **Chemical Properties**, such as combustibility and explosibility.

I. — PHYSICAL PROPERTIES.

Malleability. — Many of the metals may be reduced to thin plates, or leaves, by hammering them. Zinc is a familiar illustration, sheets of this metal being often placed under stoves to protect the floor from heat. This property is called MALLEABILITY. Gold is eminently malleable: it may be beaten into leaves so thin that a pile of eighteen hundred of them would be no thicker than a sheet of common paper.

Ductility. — Many substances may be also drawn into wire. Iron, copper, and brass wires are sufficiently familiar. The peculiar property by virtue of which they may be drawn into wire is called DUCTILITY. Glass, when heated to a bright red heat, is remarkably ductile. If a point pulled out from the mass be fastened to the circumference of a turning wheel, a uniform thread as fine as the finest silk may be wound at the rate of a thousand yards an hour.

Physical Properties. — Now fix the attention upon the fact that the wonderful malleability of gold and the surprising ductility of glass are shown *without any change in the nature of these substances*. The gold is the same material in the form of leaf as it was before it manifested its malleability. The glass in the form of thread is the identical substance which, by being drawn, manifested its ductility.

All *properties which*, like these, *a body may manifest without undergoing any change in its nature*, are called PHYSICAL PROPERTIES.

If now we examine those properties described in the early part of this section, we shall find them all to belong to this group. Extension, Impenetrability, and the rest, are properties which a body may show without any change in its nature.

II. — CHEMICAL PROPERTIES.

Chemical Properties. — Wood, by burning, shows that it is combustible. No substance can manifest the property of combustibility except by actually taking fire; and when it burns it changes to something else.

Who, not already familiar with gunpowder, would suspect it to be so violently explosive? It can show that it is explosive only by ceasing to be gunpowder, and becoming a mass of vapor. *Properties* like these, *which a body can not manifest without changing its nature*, are called CHEMICAL PROPERTIES.

This classification of properties will help us to define accurately the science whose elements we are beginning to study.

3. Natural Philosophy, or *Physics* as it is now more generally called, is the science which treats of the physical properties of matter, and of those phenomena in which there is no change in the nature of bodies.

Illustrations. — If now we look out upon the phenomena which nature presents, and will apply the test furnished by this definition, we may select, from among the multitude, those which it is the province of this science to explain. Thus, for example, we see the vapors rise; we see the raindrops fall; we listen with delight to the harmonies of music, and derive exquisite pleasure from the colors of the rainbow. In these phenomena, and in numerous others easily recognized by an attentive mind, we can detect changes in the form and place of bodies, but none whatever in their nature. We therefore expect to learn the explanation of them in the study of physics. But if we regard the more quiet yet not less imposing phenomena of the seasons, we may discover a multitude whose discussion is, by the definition, excluded from this science. The young verdure of the springtime changes at length to the matured foliage and ripening grains of summer. The fruits and hues of autumn, more somber except where enlivened by the richly colored ripening leaves of the maple or the oak, soon afterward appear, only to be in turn displaced by the snows of winter. These events are brought about by changes gradually taking place in the nature of substances; and the explanation of all such phenomena must be reserved for the science of chemistry.

SECTION II.

ON MATTER AND FORCE.

4. Matter, with respect to its Divisibility, may be considered either as Masses, Molecules, or Atoms. (*G*. 2, 3; *A*. 1, 2, 7–10.[1])

Divisibility. — Every body of matter may be cut or broken into pieces. This quality or property of matter is called DIVISIBILITY.

Examples of minute Division. — There are bodies all around us so small that we can not see them. They are in the air we breathe, and the water we drink; some of them are so minute that only the most powerful microscope can enable us to discover them. Common salt can be detected by chemical means in the air, even at great distances from the sea, and when the air is perfectly transparent.

There are living creatures so small that a million individuals together would be no larger than a mustard-seed. Yet each one must be made up of parts, else it could not move about, and take its food, as all of them can do.

A grain of cochineal dissolved in water will impart a distinct color to a whole gallon. That it may do this, it is estimated that the grain must be divided into not less than six million pieces.

Masses. — A mass is any separate portion of matter, whether large or small. A block of marble or a particle of the dust into which it may be crushed, the grain of cochineal or one of its six million pieces, is a MASS.

Molecules. — But there is a limit beyond which a mass can not be divided without changing the nature of its sub-

---

[1] *Books of reference* are indicated as follows: *G*., Ganot's Physics, by Atkinson, 8th ed., 1877; *A*., Arnott's Physics, by Bain and Taylor, 7th ed., 1877. Numbers refer to paragraphs, not to pages. The dash between two numbers means that all the paragraphs between are included: e.g., " 7-10 " refers to paragraphs 7, 8, 9, and 10.

stance. A block of ice, for example, may be crushed until its particles are like the finest dust, but, if its temperature be kept low enough while it is being crushed, every particle of the ice-powder will still be a block of ice. By applying heat, the little block is first melted, and then changed to steam, which shows that it was composed of innumerable smaller pieces. How minute must be the particles thus made absolutely invisible! Yet each one is a fragment of the original block of ice. The heat has not changed their nature. The identical particles which make up the steam composed the drop of water and the little piece of ice. But it is thought that these particles can not be divided *without changing their nature*, and they are called MOLECULES.

A molecule is a particle of matter which can not be divided without changing its chemical character.

All bodies are made up of molecules. The size of a body depends upon their number; its shape, upon their arrangement.

Atoms. — If the molecules are broken or divided, the character of the substance is changed. Let steam, for example, be passed into a red-hot gun-barrel: that which issues from the other end is not steam, but hydrogen gas, while another gas, oxygen, combines with the iron of the gun-barrel. The molecules of steam are broken into pieces, but the substance is no longer steam. These smaller parts are particles of hydrogen and oxygen.

The chemist finds that there are two pieces of hydrogen and one of oxygen obtained whenever a molecule of water is divided. But no way has yet been found to carry the division any further; and these, the smallest of all fragments into which matter can be divided, are called ATOMS. By combining together, atoms form molecules.

An atom is the smallest portion of matter that can enter into combination.

What is Inertia? — Masses of matter have no power to move themselves, nor to stop themselves when once in motion.

A heavy wheel requires force to put it in motion, or when in motion it requires force to stop it. It has no power to change its own condition. At rest, it would rest forever if left to itself; or once in motion it would forever move, unless acted upon by some force beyond itself.

Molecules and atoms also are powerless to change their own condition. It is believed that the molecules of every body are in motion always, but no molecule can in any way change its own velocity or direction. *Inertia is the property of matter which does not allow a body to change its own condition of rest or motion.*

5. Forces are either attractive or repellent. (*G.* 26.)

Force. — Bodies are sometimes in motion, and sometimes at rest. Their motion is at one time swift, at another slow. Now, inertia prevents any body from causing these changes in itself, and we must attribute them to the influence of other bodies. Whatever this influence may be, it is called FORCE.

Force is the name given to the influence which tends to produce or diminish or in any way change the motion of bodies.

Attraction. — When a body is not supported it falls to the ground. This familiar event illustrates the tendency of bodies to approach each other. Moreover, we have seen that bodies are composed of molecules, so small that the most powerful microscope can not reveal them, yet we must think of each as a separate body as truly as though the eye could measure its diameter. Now, by what influence are they held together? It is doubtless the same invisible force by which a body is drawn to the earth when not supported. It is a fact that all bodies, however large or small, have a tendency to approach each other. The *force* which causes this tendency is called ATTRACTION.

Repulsion. — If a ball of India-rubber be pressed in the hand, it is made smaller; its molecules are brought nearer together. When the pressure is removed, they instantly spring to their former position. While springing back the molecules are evidently being thrust away from each other.

Or try the following experiment (Fig. 2): Suspend a pith-ball, or a little ball of cotton, by a fine silk thread: briskly rub a warm dry strip of glass with a woolen cloth: bring the ball and glass together for a moment, after which it will be found that the ball will fly away from the glass, and show so strong an aversion to it that they can not be brought together. That which keeps them apart is called REPULSION.

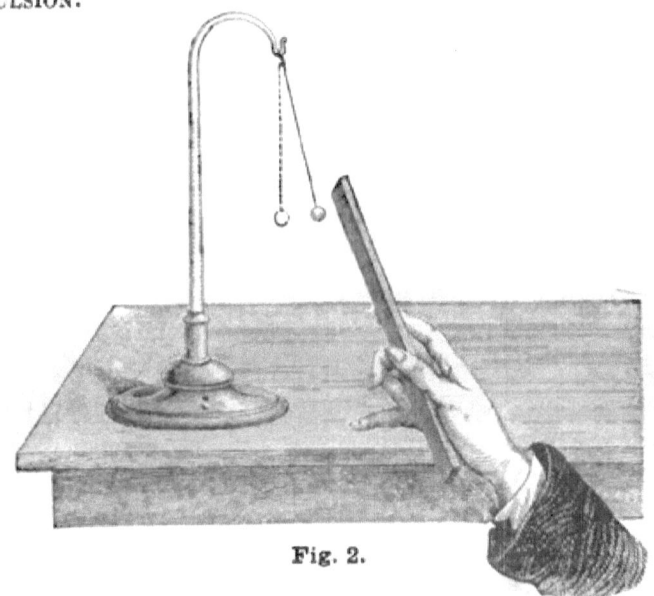

Fig. 2.

An influence under which bodies tend to separate is called REPULSION.

Molecular Repulsion. — The action of repulsion among molecules is more universal than among masses. It is illustrated by many familiar facts. If, for example, a bladder be filled with cold air, and then heated, it will burst. Repulsion drives the molecules of air apart, and pushes them through the bladder. When a drop of water is heated it becomes steam, and fills a space about seventeen hundred times larger than before.

Notice that in these examples the separation of the molecules is due to heat. *It is believed that all molecular repulsion is the same influence as that which shows itself as* HEAT.

The Forces of Nature. — We speak of the *forces of nature*, and call them wind, water, gravitation. This is well, because these names have been given to familiar forms of force. We will continue to use these terms: at the same time, let us do justice to the simplicity of God's stupendous works, by remembering that all the forces of nature are only different manifestations of *attraction* and *repulsion*.

6. Attraction receives different names according to the circumstances under which it acts. Gravitation, Cohesion, Adhesion, and Chemism are its most common forms. (*G.* 84–87; *A.* 18–28, 86, 87.)

I. — GRAVITATION.

Gravitation defined. — Gravitation is that form of attraction which is exerted upon all bodies, and throughout all distances. The leaf, the fruit, the snow-flake, fall to the ground because they are attracted thither by gravitation. They press upon its surface because the same force continues to act after they reach the earth. No distance can outreach it, for it is the bond which holds the heavenly bodies in their orbits. Nor can any substance cut it off, or even diminish its action; for, if the earth should come between the sun and moon, these two bodies would still attract each other with the same degree of force.

Governed by two Laws. — The intensity of gravitation between bodies depends on two things, — the quantity of matter contained in them, and the distance between their centers. The "law of gravitation," discovered by Newton, consists of two parts which express these two relations. We may study them as two laws.

The First Law of Gravitation. — Suppose two bodies, one containing twice as much matter as the other, to attract a third. The force exerted by the first will be twice as great as that by the other. If one body weighs nine tons and another three tons, then a third body equally distant from them will receive three times as much attraction from the first as from the second. These facts illustrate the law that *the force of gravitation varies directly as the quantity of matter exerting it.*

The Second Law of Gravitation. — Suppose a body to be *twice as far* from the center, or source of attraction, at one time as at another. In the first position, the attraction will be only *one-fourth* as strong as in the second. If the distance be three times as great, the force will be one-ninth as strong. If two distances are as 3 : 4, the attractions will be to each other as 16 : 9. These facts illustrate the law that *the force of gravitation varies inversely as the square of the distance through which it acts.*

Weight. — The weight of a body is the downward pressure which it is able to exert. It is due to the attraction of gravitation exerted by the earth. Weight must therefore increase or diminish in exact accordance with the laws of gravitation.

Is proportional to the Quantity of Matter. — According to the first law, a double quantity of matter is attracted by the earth with a double force, and it therefore weighs twice as much. For the same reason, if the quantities of matter in two bodies are as 1 : 4, their weights must also be as 1 : 4. That is: *the weights of bodies are proportional to the quantities of matter in them.*

And inversely proportional to the Square of the Distance from the Earth's Center. — According to the second law, the greater the distance from the earth, the less will a body weigh. Now, distance from the earth is measured *from its center*. When on the surface of the earth, a body is four thousand miles from the center; suppose it were possible to carry the body to a height of four thousand miles above the surface, its distance from the center would be *doubled*, and its weight would be reduced to *one-fourth*.

Not limited to Bodies on the Earth. — Since the laws of gravitation are universal, a body on the moon or any other heavenly body has weight.

Not the same as Mass. — By mass, we mean the *quantity of matter* in a body. We speak of the body itself as *a mass* to distinguish it from molecules and atoms; but *the mass of a body is the quantity of matter it contains.*

A body on the moon, for example, would have the same quantity of matter in it as if upon the earth. Yet the moon could exert much less attraction than the earth, and hence the *weight* of the body would be less. Mass is the quantity of matter in a body; weight is the force of attraction exerted on it. The mass of a body is everywhere the same; its weight is different at different places. But *at any one place the weights of all bodies are proportional to their masses;* and for this reason we can compare the quantities of matter in bodies by weighing them.

II.—COHESION.

Cohesion defined. — Cohesive attraction holds the molecules of a body together, and enables it to keep its form and size. A cubical block of wood remains a cube only because its molecules are held together by this force. Were it not for its action, all bodies would at once dissolve into their ultimate molecules, and vanish.

COHESION *is that form of attraction which acts between the molecules of the same body.*

Its Power is very great. — The strength of cohesion is often very great. The molecules of a piece of iron are so strongly bound by it, that a weight of five hundred pounds may be lifted by means of a wire one-tenth of an inch in diameter. Even a strip of paper is not easily broken by a force acting exactly in the direction of its length.

But it acts through Insensible Distances. — The distance through which cohesion can act is quite too small to be measured. Let the parts of a body be separated, and the strength of the giant is gone.

When a body is broken, its parts can be made to cohere again only with great difficulty. In a few soft bodies, like wax, a slight pressure will force the molecules near enough together for cohesion to take hold of them; in others, the pressure required is much greater; while in the majority of substances it is so great as to be practically impossible.

Welding. — The smith unites two pieces of iron by *welding*. He softens the iron by heat, then puts the two pieces together, and unites them by the heavy blows of his sledge. All that he does is simply to push the yielding molecules of the two pieces of iron into very close contact; this done, cohesion grasps them, and the two pieces become one.

III. — ADHESION.

Adhesion defined. — If the hand be plunged into water, it comes out covered with a thin film of the fluid; it may be immersed in alcohol with the same effect. When one writes upon the blackboard, he leaves fine particles of the crayon clinging to the surface of the board.

In all such cases we notice that there is an attraction between particles of different kinds of matter. Attraction between particles of unlike kinds is called ADHESION.

By what Experiment may we illustrate it? — A very pretty experiment is shown in Fig. 3. It illustrates the adhesion between water and brass. A round plate of brass, having a handle fastened to its center, is laid flat upon the surface of water, and then slowly and gently lifted. The water under it is also lifted a little, as the picture shows it to be.

Fig. 3.

You can use a plate of wood or of glass in the same way.

Adhesion of Solids to Solids. — The value of glue and

cement is due to the powerful adhesion which acts between them and the surfaces of solid bodies which they bind together.

Adhesion of Liquids to Solids. — If small glass tubes be inserted in a vessel of water, it will be seen that the fluid instantly springs upward, and remains at rest in the tubes considerably above its general level. (See Fig. 4.) Along the outside surface of the tubes the water also climbs to a little height.

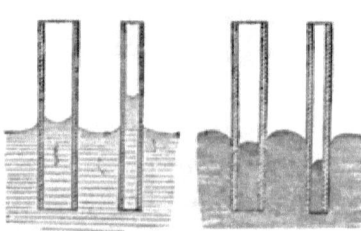

Fig. 4.   Fig. 5.

If tubes be inserted in a vessel of mercury, this fluid will be pushed down. (Fig. 5.) The mercury inside the tubes will be considerably below the general level, while the fluid against the outside is also depressed. Here are two well-marked cases of the action of adhesion.

Whenever a piece of glass is plunged into water it comes out wet, but when plunged into mercury it comes out as free from the liquid as when it entered; and by repeated experiments it is shown that all liquids which will wet the sides of the tube will be *lifted*, while those which will not will be *pushed down*.

Capillarity. — The elevation or depression of fluids in small tubes is called CAPILLARITY. The attraction causing it is, however, nothing different from adhesion.

It causes Liquids to penetrate Porous Solids. — An easy experiment strikingly illustrates this action. Take a common bottle, eight or ten inches high, and wrap it in a sheet of white blotting-paper, whose edges must be secured by a bit of wax. Place the bottle, now prepared, upon a dinner-plate. Pour water upon the plate to cover the lower edge of the paper; and immediately the fluid will be seen rapidly climbing the sides of the bottle, which it will not cease to do until it has reached the top.

Explanation. — The rise of the water is due to the attraction between its particles and those of the paper and glass. This force, acting downward from each particle of the paper through the definite but imperceptible distance to the one below it, lifts a particle of water. The next particle of paper above then lifts it higher. Indeed, the successive particles of paper upward are the successive steps of a ladder, up which the water is impelled by capillary force.

Familiar Examples. — Numerous familiar facts are explained by this experiment. Oil is carried up the lamp-wick to supply the flame with fuel. By a similar action, water is distributed through loose soils to keep them moist and fertile. So, too, in a great degree, the sap of plants and trees is carried to their summits; and even in the animal system the circulation of blood through the minute blood-vessels is materially aided by capillary action.

The Law. — In Fig. 4, the water is represented as being lifted to different heights in the tubes. The height to which any fluid rises depends upon the size of the tube. If the diameter of one tube be just one-half that of another, water will invariably rise in it twice as far. If the diameters of two tubes have the ratio of 4 : 3, then the water will rise in them to heights whose ratio is 3 : 4. These facts illustrate the law that *the heights to which a liquid rises in different tubes of the same material are inversely proportional to the diameters of the tubes.*

A tube one-hundredth of an inch in diameter will lift water to a height of about four inches.

IV. — CHEMISM.

Chemism defined. — When coal burns, the carbon of which it is composed *combines* with the oxygen of the air. They form a new substance called in chemistry *carbonic dioxide*. The properties of carbonic dioxide are very unlike those of either coal or oxygen.

This action occurs among the *atoms* of coal and oxygen.

Under the influence of the heat these unlike atoms are drawn together, and then so firmly held that they seem to be one body. They form *molecules* of carbonic dioxide.

The attraction which holds the atoms together in a molecule is called CHEMISM. Chemical affinity is another and an older name for it.

SECTION III.

ON THE FUNDAMENTAL IDEAS.

7. The fundamental ideas in regard to matter and force are suggested by the words, Molecules, Inertia, Attraction, and Repulsion.

These terms have been already defined; but they may suggest to the mind much more of what is believed about the minute constitution of matter than their definitions contain.

Principle suggested by the Term *Molecule.* — Every body of matter is made up of a multitude of little particles, which do not touch each other, which are forever in motion, and which can not be divided without changing their nature.

Principle suggested by the Term *Inertia.* — No body of matter — a mass, a molecule, nor an atom — has power to change its own condition of rest or motion.

Principle suggested by the Term *Attraction.* — All bodies of matter — masses, molecules, and atoms — tend to approach one another.

Principle suggested by the Term *Repulsion.* — There is also an influence under which molecules tend to separate, and which forbids their actual contact in a body.

Of these four Ideas. — These four ideas may be so combined as to yield the explanation of almost all the phenomena which appear in nature. A great city, with all its various forms of architecture and machinery, is built of a few familiar substances, such as wood and iron and stone. This fact may excite our admiration of the intelligence and skill of man. What, then, must be our feelings when we discover that these four simple ideas are the elements out of which the sublime fabric of the universe has arisen! The whole system of material things is simple and orderly, displaying the infinite knowledge, power, and skill of a divine Architect.

SECTION IV.

REVIEW.

I.—SUMMARY OF PRINCIPLES.

In the foregoing chapter we have found that:—

All properties of matter are either physical or chemical.

Physics or Natural Philosophy is the science which treats of the *physical* properties of matter, and of those phenomena in which there is no change in the nature of substances.

Matter may be considered as consisting of masses, molecules, and atoms.

Forces are either attractive or repellent.

Attraction is called

Gravitation, when it acts between masses;
Cohesion or adhesion, when it acts between molecules;
Chemism, when it acts between atoms.

The force of gravitation varies directly as the quantity of matter exerting it, and inversely as the square of the distance through which it acts.

Volume is the space occupied by a body.

Mass is the quantity of matter in a body.

Weight is the pressure of a body downward, caused by the attraction of the earth.

The fundamental ideas of matter and force are suggested by the words, Molecule, Inertia, Attraction, and Repulsion.

II.—SUMMARY OF TOPICS.

1. Properties of matter. — Extension. — Impenetrability. — Indestructibility. — Elasticity.

2. Malleability. — Ductility. — Definition of physical properties. — Definition of chemical properties.

3. Definition of Physics. — Illustrations.

4. Divisibility. — Examples. — Masses. — Molecules. — Atoms. — Inertia.

5. Force. — Attraction. — Repulsion. — Molecular repulsion.

6. *Gravitation* described. — Governed by two laws. — The first law. — The second law. — *Weight* defined. — It is proportional to quantity of matter. — Inversely proportional to square of the distance from the earth. — Not limited to bodies on the earth. — Not the same as mass. — *Cohesion* defined. — Its power. — At insensible distances. — Welding. — *Adhesion* defined. — Experiment. — Of solids to solids. — Of liquids to solids. — Capillarity. — Familiar examples. — The law. — *Chemism* defined.

7. The fundamental idea suggested by the term Molecule. — By the term Inertia. — By the term Attraction. — By the term Repulsion.

III.—PROBLEMS.

1. With how many times greater force will a body be attracted by a mass of iron weighing 9 tons, than by a block of stone weighing 3 tons, when both are at the same distance from it? *Ans.* 3.

2. Two lead balls, one weighing 5 ounces and the other 12 ounces, are hanging at a distance of 10 feet from a third; what relative degrees of force do they exert upon it?

3. One ball of lead attracts another through a distance of 10 feet with a force of 8 pounds; what force would it exert if placed at a distance of 20 feet? *Ans.* 2 pounds.

4. A body is at one time 50 feet, and at another 75 feet,

from a mass of rock; what are the relative forces exerted upon it in the two positions? *Ans.* 9 : 4.

5. Two bodies, one weighing 6 pounds, and the other 9 pounds, are attracting a third. The first is at a distance of 25 feet, the second of 50 feet; what relative attractions do they exert?

$$\frac{6}{25^2} : \frac{9}{50^2},$$ which, reduced, is the *Ans.* 8 : 3.

6. At the surface of the earth a body weighs 10 pounds; what would it weigh if carried to a height of 5 miles above the surface? *Ans.* 9.975 pounds.

$$(4005 \text{ miles})^2 : (4000 \text{ miles})^2 :: 10 \text{ pounds} : x.$$
$$x = 9.975 \text{ pounds}.$$

CHAPTER II.

THE THREE PHYSICAL FORMS OF MATTER.

SECTION I.

APPLICATION OF THE FUNDAMENTAL IDEAS.

8. ATTRACTION and REPULSION acting upon the molecules of bodies produce the three physical forms of matter, — Solid, Liquid, and Gaseous.

The Action of the Molecular Forces. — Between the molecules of every body, two sets of forces, attractive and repellent, are continually struggling. Just in proportion as one or the other prevails, the body will be a solid, a liquid, or a gas. In a solid body attraction prevails, and its molecules are firmly bound together. In a liquid body the attraction is almost equaled by the repulsion, and the molecules are left free to move easily among themselves. In a gaseous body the repulsion exceeds the attraction, and the molecules are driven away from each other to the greatest possible distance. The solid rock, the mobile water, and the rushing air are types of these three grand divisions to which all bodies belong.

Changes of Condition in Nature. — Numerous and familiar changes of form are due to the action of heat. Ice, for example, when heated, becomes water; and water, when heated still more, rises in vapor to form the floating clouds. In this case molecular repulsion is increased until it is stronger than molecular attraction. Or, suppose the action to be reversed. Imparting their heat to other bodies, the clouds are changed to water, and water again to solid ice and feathery snow. In this case molecular repulsion is diminished until it is weaker than attraction.

Artificial Changes. — Imitating nature, we may to a limited extent, by the use of heat, change the form of various bodies; and numerous arts of life spring from the application of this power. By the repulsive force of heat, the metallic ores are melted, and the useful metals obtained. By the same force iron is liquefied, that it may be molded into requisite forms of strength, of beauty, or of use, demanded in the arts. The expansive force of steam is but the repulsive force of heat.

SECTION II.

ON THE CHARACTERISTIC PROPERTIES OF SOLIDS.

9. The characteristic properties of solid bodies are Hardness, Tenacity, Malleability, Ductility, and Crystalline Form. (G. 93–95; A. 59–64.)

Hardness. — The particles of solid bodies are held together by cohesion, much more firmly in some than in others. Those in which they are held with the greatest force will most successfully resist the pressure of others. By the term HARDNESS, we refer to that property of solids which enables them to resist any action which tends to wear or scratch their particles away.

Hardness does not imply Strength. — A piece of glass will scratch an iron hammer, which proves it to be harder than iron; yet glass is very fragile, easily broken by the stroke of soft wood, indeed, by almost any thing that can inflict a blow.

Neither does Hardness imply Density. — The diamond is the hardest of substances, while gold is so soft as to be easily cut with a knife; yet gold is four times as dense as the diamond. Mercury is a fluid, and, of course, has no hardness; yet it is nearly twice as dense as the hardest steel.

Tempering. — The process called tempering, or annealing, consists in regulating the hardness of a body by the action of heat. Steel, when in its hardest condition, is too brittle to be used in the arts; but by heating it to a temperature determined by the use to be made of it, and then slowly cooling it, the steel may receive any degree of hardness desirable. It may be made almost as soft as soft iron, or it may become nearly as hard as the diamond.

Tenacity. — When a rod of iron is stretched in the direction of its length, it will be found that great force is required to pull it apart. The property, in virtue of which a body resists a force which tends to pull its parts asunder, is called TENACITY.

Of Metals. — The metals are more tenacious than other solids; and among metals, iron in the form of cast-steel stands at the head of the list. A rod of cast-steel, the end of which has an area of one square inch, will support a weight of 134,256 pounds.

It has been found by experiment that the tenacity of a bar is in proportion to the area of its cross-section, and entirely independent of its length.

It has also been shown that the tenacity of a metal is greatly increased by drawing it into wire. The cables of suspension bridges are, for this reason, made of fine iron wire twisted together.

Malleability. — The particles of many solid bodies may be displaced without overcoming their cohesion. By the blows of a hammer, the molecules of many metals may be shifted about, without breaking them apart, until the bodies are reduced to the form of thin plates or leaves. By passing the metal between the rollers of a rolling-mill, the great pressure exerted will produce the same effect. This property, in virtue of which a body may be hammered or rolled out into thin leaves or plates, is called MALLEABILITY.

Of Metals. — This property is possessed in a high degree by many of the metals. Under the hammer, lead is the most malleable of the useful metals; tin stands second, and gold third, on the list. In the rolling-mill, gold is the most malleable, silver is second, copper third, while tin stands in the fourth place on the list. (*Silliman's Physics*, p. 138.)

Ductility. — If, instead of being reduced to thin plates, the substance may be drawn into wire, the property thus shown is called DUCTILITY. This property is closely allied to malleability, but metals do not possess both in an equal degree. Platinum, for example, which is seventh on the list of malleable metals, stands first on the list of those which are ductile. This metal has been drawn into wire finer than a spider's thread.

Crystalline Form. — The attraction among the molecules has not brought them together at random, nor in disorder. A flake of snow, when seen through a microscope, is found to be as symmetrically formed as a swan's feather; and water frozen on the window-panes in winter shows a beautiful variety of tree-like forms. These definite and regular forms in which solid substances occur are called CRYSTALS; and any process by which they may be obtained is called a process of CRYSTALLIZATION.

In Nature. — In the formation of solid bodies their tendency to take a crystalline form is almost universal. The same substance generally takes the same form, but in different substances the shapes of crystals may be wonderfully unlike. Lead, in its most common ore, called *galena*, is found crystallized in cubes. Specimens of these cubes are often found as perfect as could be chiseled by an artist.

But the larger number of solid bodies around us do not appear to have these definite crystalline forms. They have been made solid under circumstances which did not allow the molecular forces to act freely. In many cases, however, if we break open a body whose external form is not regular, we may discover that it is, after all, a crystallized body, by noticing that it is made up of multitudes of small crystals, very closely packed together. This is true of many rocks.

Made artificially. — Even when no indication of a crystalline structure can be seen, the substance can often be made to assume it by some artificial process. The best method is to dissolve the solid in water or some other liquid, and allow the solution to stand in a quiet place where it may evaporate slowly. Common salt and alum are substances which readily and beautifully illustrate this process. The more slowly the water evaporates, the more perfect will the crystals be.

SECTION III.

ON THE CHARACTERISTIC PROPERTIES OF LIQUIDS.

10. Liquids have elasticity and some other properties in common with solid bodies. But, since the attraction and repulsion among their molecules are very nearly equal, we find that Mobility is their characteristic property. ($G.$ 97, 98; $A.$ 74.)

Elasticity. — When submitted to pressure, liquids are compressed; and when the pressure is removed they instantly spring back to their original volume: they are *elastic*.

The force with which a liquid springs back to its former size, after being compressed, is exactly equal to the force which compressed it: it is, for this reason, said to be *perfectly elastic*.

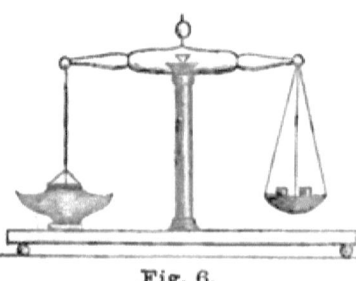
Fig. 6.

Compressibility. — It requires a very great force, however, to compress a liquid in the least degree; so great, that, until improved means of experiment were contrived, liquids were thought to be incompressible. Water at a freezing temperature, when pressed by a force of fifteen pounds to the square inch, is condensed only .0000503 of its volume.

Attraction and Repulsion nearly equal. — That the attractive and repulsive forces among the molecules of a liquid are not exactly equal, may be shown by a pretty experiment.

To one end of a scale-beam (Fig. 6) a disk of brass is suspended, and accurately balanced by weights in the opposite scale-pan. Then let the disk be brought to rest upon the surface of water in a vessel, and it will be held there with considerable force. If the disk be two inches in diameter, weights equal to two hundred grains may be placed upon the opposite pan before it will be torn from the water. But we notice that a film of water still adheres to the disk, having been torn away from the water beneath it. The weight of two hundred grains has simply overcome the *cohesion* of the water. We thus learn that the attraction is a trifle stronger than the repulsion.

Mobility. — But the attractive and repulsive forces are *nearly* balanced; and, if we now remember that water consists of molecules, it is not more difficult to see that there must be freedom of motion among them, than it is to see that shot, or smooth balls of any other kind, will roll easily upon each other.

The molecules of water are balls infinitely smaller than shot, but, while the most powerful microscope fails to reveal them, the mind can see them, so small, so round and smooth, that they roll and glide among themselves with the greatest freedom.

This *freedom of motion among the molecules of a substance is called* Mobility.

Mobility is the property of a liquid which makes it differ from a solid. Destroy the *mobility* of water, and it becomes a solid, — ice. Impart *mobility* to a solid, and it ceases to be a solid; it becomes a liquid, as when ice is melted.

SECTION IV.

ON THE PRESSURE OF LIQUIDS.

11. At any point inside of a body of liquid at rest, there is equal pressure from all directions.

Hence a fluid will rest only when its upper surface is level; but the level surface of a large body of water is convex. (*G.* 105, 108, 110; *A.* 304.)

Liquids press in all Directions. — In order to see that, because the particles of a liquid are free to move, they must be exerting pressure in all directions, we will suppose a number of very smooth balls to be arranged as in Fig. 7. The weight of the ball A will be a downward pressure upon the balls B and C. These, being free to move, will be pushed aside. The ball B, moving toward the left, will push between the balls D and E, while the ball E, moving upward, will exert an upward pressure. Just so the small molecules of a liquid are exerting pressure downward, upward, and laterally. This pressure is due to gravitation.

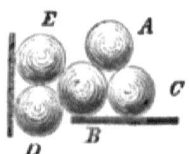

Fig. 7.

Moreover, the pressure of the liquid must be exerted

Equally in all Directions. — For, if the pressure at any point were greater in one direction than another, the liquid would *move* in that direction on account of its mobility. If, then, the liquid is *at rest*, the pressures in all directions must be just balanced.

Experimental Illustration. — An experiment may help to illustrate this principle. If a disk of metal be held in the middle of a jar of water, it is easy to see that it must be pressed downward by the weight of the water just above it; but it may not be so clear that it is pushed upward by an equal force. Taking a lamp-chimney, and putting the string

NATURAL PHILOSOPHY. 27

handle of the disk (Fig. 8) through it, hold the disk tightly against the lower end of the tube until it is pushed down to the middle of the water. Now drop the string: the heavy disk does not sink, but remains tightly pressed *upward* against the tube by the water. If water be allowed to enter the tube, it will press down upon the disk; and, when it has filled the tube almost to a level with the water outside, then the disk falls, suggesting that the upward and downward pressures are equal.

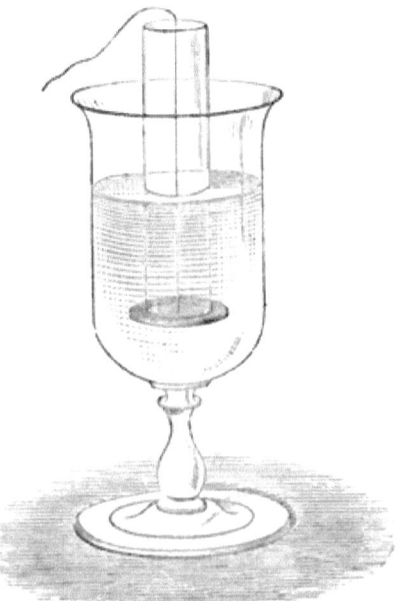

Fig. 8.

Another Illustration. —Numerous simple experiments might be given to illustrate this important principle: one other must suffice. Glass is eminently brittle. It may be blown into sheets as thin as the finest paper-cambric. In this condition, the weight of a few grains resting upon it in the air would crush it. Yet, placed near the bottom of the deepest cistern, it will support the weight of all the water above it, and remain unbroken. This could not happen if the pressure of the water upon it were not equal from all directions.

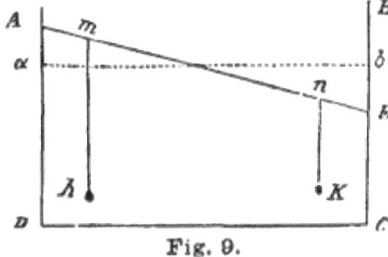

Fig. 9.

The Surface of Water at Rest is level. — The truth of this principle may be seen by attentively examining Fig. 9, which represents a section of a vessel containing water, the surface of which has for the moment been thrown into the position indicated by the line A E. Refer to any two points in the water, as h and K. We see that the downward pressure at h would be the weight of the water above that point, — a column $m\ h$. But the pressure at that point is equal in *all* directions, so that the water between h and K would be pressed toward K by a force equal to the weight of the column $m\ h$. In just the same way we may show that at K the water is being pressed toward h by a force equal to the weight of the column n K. The column $m\ h$ is greater than n K; and, since the water is *free to move*, it will yield to the greater pressure, and go toward K until the two forces are equal. The two forces will be equal only when m and n are in the same level surface, $a\ b$.

But a level Surface is convex. — The surface of water will be at rest when the force of gravitation acts upon all points of it alike. That the attraction of the earth may be equal on all points, they must be equally distant from the center of the earth : to be at the same distance from the center of the earth, they must form a curved surface. In case of large bodies of water, of the oceans for example, the convexity can be seen. It is shown by the ancient observation that the topmast of an approaching ship is the part first seen from port.

12. Since the surface of water at rest must be level, we infer that water confined in pipes or close channels will always rise in them as high as the source from which it comes.

Upon this principle cities are often supplied with water.

The same principle explains the phenomena of springs and artesian wells.

Water in Pipes will rise as high as its Source. — No matter what may be the shape of the vessel, the surface of the liquid it holds must be just as high in one part of it as in another. Moreover, a pipe leading from a vessel is a part of the vessel which holds the water.

The vessel shown in Fig. 10 has a very irregular shape. There is first the large vase at the left hand, then the horizontal tube, and finally the tubes reaching upward from the last; yet it is all one vessel, because the water can pass freely from one part to another.

If water is poured into the vase, it will rise just as fast in the tubes, and will at last stand at the same height in all parts, as the picture shows it to be.

The Supply of Water to Cities. — A pipe leading from a reservoir of water on a hill outside a city may be buried

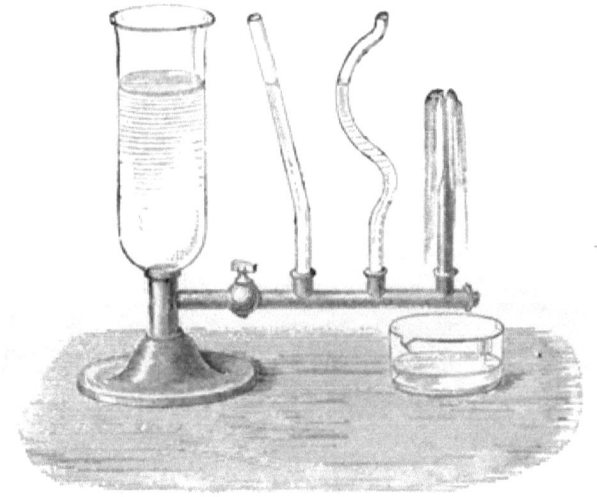

Fig. 10.

in the ground, passed down the hillside and through the streets, and be provided with branches leading into cisterns in every dwelling. Unless these cisterns are higher than the water in the distant reservoir, the water will flow down the hillside, through the streets, and up the branches, into the dwellings, and supply them all with water. Many cities are in this way conveniently supplied with abundance of water, not only for private dwellings, but also for public fountains and manufacturing purposes.

Springs. — The rocks which compose the earth are arranged in layers, called strata, which are generally more or less oblique as represented in Fig. 11. Some of these strata will allow water to soak through them: others will not. In the figure the dotted portions $a\,a\,a$ indicate the porous strata.

Now, water falling on the surface of the earth at c will settle through the loose or porous material until it reaches the rock, which it can not penetrate. Flowing along the surface of this rock, it will issue from the hillside at S, and thus form a *spring*.

Artesian Wells.—Again, the water, falling upon the

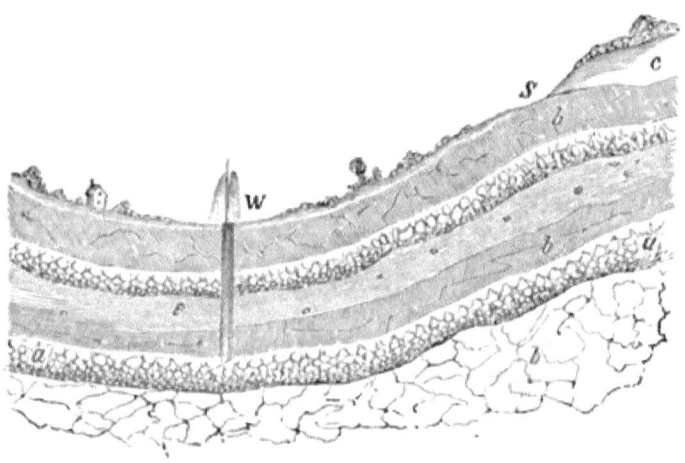

Fig. 11.

surface, and passing through other porous layers, at length comes in contact with a rock which it can not penetrate, and flows along its surface. The basin-shaped part, $a\,a$, of the porous layer, would thus in time become filled with water; indeed, the entire layer reaching to the surface of the earth in both directions might thus be filled. If, then, a well as W be sunk through the mass down to this saturated layer, the water will collect in the well, and rise sometimes to the surface of the ground above, and sometimes it even spouts in jets many feet above the surface.

Such wells are often bored to very great depths, and are

called artesian wells. One of these wells was bored in Louisville, Ky., to the depth of 2,086 feet. Another in St. Louis has a depth of 2,199 feet. The supply of water furnished is often very abundant. The famous Grenelle well, in Paris, yields daily 600,000 gallons.

13. The pressure of a liquid on the bottom of the vessel which holds it is independent of the shape of the vessel.

It depends on the depth of the liquid, and the area of the bottom of the vessel.

It equals the weight of a column whose base is the base of the vessel, and whose height is the depth of the liquid. (*G.* 104; *A.* 303.)

The Pressure is independent of the Shape of the Vessel.— This may be proved by experiment. The essential parts of an apparatus for this purpose are represented in Fig. 12. A glass tube, A B, bent twice at right angles, contains mercury. The height of mercury in one arm is shown by a graduated scale, and to the other arm vessels of various forms and heights may be attached. When a vessel, G, is filled with water,

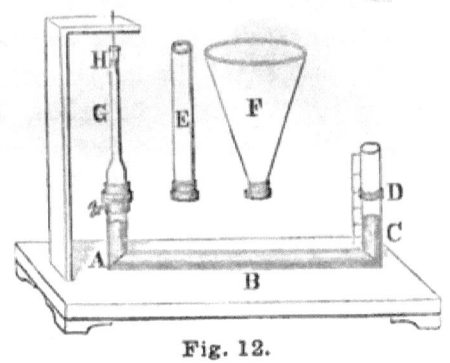

Fig. 12.

the fluid presses upon the mercury at A, and pushes it up in the arm C D; the height to which it rises being shown by the graduated scale. Now let the vessel be removed, and another, in the form shown at E, be put in its place. If water be poured into this vessel until it stands as high as it did in the other, the mercury will be seen to rise in C D to the same point as before. Vessels of various other forms may be used; but, if all are of the same height, the water which fills them will push the mercury to the same point on the scale. We infer that the pressure of a fluid downward is quite independent of the shape of the vessel and of the quantity of fluid it contains.

The Pressure depends on the Depth of the Liquid. — If a tube twice as high as the vessel E, in Fig. 12, be used, and filled with water, the mercury will be seen to rise just twice as far as when the other vessels were employed; and by repeated experiment it is found that the pressure is in proportion to the height of the column of water which exerts it.

The Pressure depends also on the Area of Base. — In these experiments the area receiving the pressure at the bottom of the vessels is the same for all. If next we suppose the bottom of the vessel to be made just twice as large, there would be just twice as much water resting upon it, and therefore just twice the downward pressure. In any case, the pressure is in proportion to the area of the base of the vessel.

To calculate the Pressure. — If the pressure depends only on the size of the base and the height of the column, then it must equal the *weight* of a column whose base is the base of the vessel, and whose height is the depth of the liquid. Now, *one cubic foot of water weighs* $62\frac{1}{2}$ pounds; and, if the number of cubic feet of water which exerts the pressure be multiplied by $62\frac{1}{2}$, the amount of pressure in pounds will be obtained.

Illustration. — Thus, suppose a vessel, represented by E F C D in Fig. 13, to be full of water: the pressure on its bottom is the weight of the column A B C D. Let the area of the bottom be three square feet, and the depth of the water be eight feet; then the pressure is the weight of 24 cubic feet of water, and $24 \times 62\frac{1}{2}$ pounds, or 1,500 pounds, is the pressure exerted.

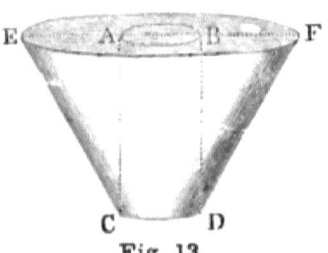
Fig. 13.

Since the pressure is equal in all directions, we may

obtain the amount of *pressure against any portion of surface*, either in the bottom or sides of the vessel, by finding the weight of a column of water whose base is the surface pressed upon, and whose height is the depth of the water *to the middle point* of that surface.

For example: suppose we would know how much pressure is borne by one square foot of the *side* of a vessel at a depth of ten feet below the surface of the water. We must understand that ten feet is the distance from the top of the water to the *middle point* of the square foot; then the pressure will be the weight of a column of water whose base is one square foot and whose height is ten feet. Such a column will contain ten cubic feet of water, and its weight will be $10 \times 62\frac{1}{2}$ pounds.

14. A solid body, when immersed in a fluid, is pushed upward by it with a force equal to the weight of the fluid it displaces.

It follows from this principle: —

1st, That a solid body, lighter than water, will sink far enough to displace a quantity of water just equal to its own weight.

2d, That a solid body, heavier than water, will weigh less in water than in air, the difference being the weight of the water displaced by it. (*G.* 116, 119; *A.* 320, 325–327.)

Solid Bodies in Water are pressed upward. — If, for example, a piece of wood be pushed down into a vessel of water, we find it struggling to rise to the surface: it is pressed upward by the water under it, and considerable force of the hand is required to keep it down. Or, if a stone be suspended in water, it is lighter than when in air: the water under it pushes upward against it, and thus supports a part of its weight.

With Force equal to the Weight of Water displaced. — Let us suppose that a block of marble is suspended in a vessel of water (Fig. 14). The upward pressure against its lower surface, $a\,b$, is equal to the downward pressure of the water at that depth; and this downward pressure is equal to the weight of the column of water, $e\,f\,b\,a$.

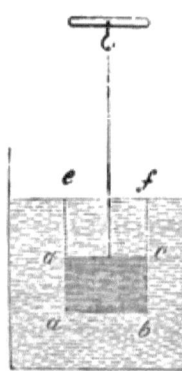

Fig. 14.

Now, a part of this upward pressure sustains the column of water, $e\,f\,c\,d$, and the rest of it is exerted upon the marble. To sustain the column, $e\,f\,c\,d$ requires an upward pressure equal to its weight; and hence there is left a pressure against the surface $a\,b$, equal to the weight of a column of water, $a\,b\,c\,d$, but this water is displaced by the marble. The upward pressure against the block is therefore equal to the weight of the fluid displaced.

The Principle of Archimedes. — We owe the discovery of this important principle to Archimedes, one of the most eminent philosophers of antiquity; and to this day it is called the PRINCIPLE OF ARCHIMEDES. Its applications are numerous. It helps the chemist to distinguish one substance from another, and the merchant, often, to judge of the purity and value of his merchandise. In any case it enables the inquirer to determine the size or volume of a solid body, however irregular; and it has, moreover, led to valuable improvements in marine architecture and in other arts.

If the Solid is lighter than Water. — The weight of the water displaced by a block of wood will just equal the weight of the wood itself. A pound of wood will displace a pound of water. But a pound of wood is larger than a pound of water, so that only part of the wood will be immersed. A tin basin and a wooden bowl of the same size will displace an equal volume of water, if the walls of the basin are thin enough so that the two bodies have the *same weight*. Upon this principle iron ships are built. An iron ship will sink no farther than one of wood of the same size, provided the walls of iron are so thin that the two ships shall be of the same weight.

NATURAL PHILOSOPHY. 35

If the Solid is heavier than Water. — If a solid is heavier than water, the upward pressure of the fluid can support only a part of its weight. The weight supported will be equal to the weight of the water which the solid displaces. Thus, for example, a piece of marble which weighs 10 ounces in air will be found to weigh only 6.3 ounces in water. The upward pressure of the water is equal to the difference, 3.7 ounces; and this is the weight of the water which the marble displaces, and whose bulk is, of course, just equal to the bulk of the marble.

15. The specific gravity of a substance is its weight compared to the weight of an equal volume of some other body taken as a standard.

To obtain it, different methods must be taken, according as the body is a *gas*, a *liquid*, or a *solid*. (*G.* 123, 125, 130.)

Specific Gravity. — The specific gravity of a substance shows how many times heavier it is than an equal volume of some other body. The standards used are *water* and *air;* water for all solid and liquid bodies, and air for all gases. In chemistry *hydrogen* instead of air is the standard for gases. Then, when we say, for instance, that the specific gravity of gold is 19, we only mean that a cubic inch of gold will weigh 19 times as much as a cubic inch of water. The specific gravity of oxygen gas is 1.106: that is to say, a cubic inch of oxygen gas will weigh 1.106 as much as a cubic inch of air. Or, compared with hydrogen, its specific gravity is 16: that is to say, a cubic inch of oxygen weighs 16 times as much as a cubic inch of hydrogen. The following simple rule must evidently cover all cases of finding specific gravity: *Divide the weight of the body by the weight of an equal volume of the standard.*

How shall these two Weights be found? — The weight of the body whose specific gravity is sought can be found directly with the balance, but the weight of an *equal volume* of the standard can not. It is not easy to know the exact volume of the body, which is often very irregular in shape. Different methods of experiment must be used in different cases.

I. — OF GASES.

How may the Specific Gravity of a Gas be found? — To obtain the specific gravity of a gas, divide the weight of a convenient portion of it by the weight of an equal portion of air or hydrogen.

To get the weight of equal portions of gases, is, however, a difficult process, requiring many precautions. Without trying to give the details of the operation, we may describe it in general terms. A glass globe is first weighed when full of air (Fig. 15). The air is then taken out of it by means of an air-pump, and the globe is again weighed. The difference in these weights is the weight of the globe-full of air. The globe is then filled with the gas whose specific gravity is desired, and again weighed. The difference between this weight and that of the empty globe is the weight of the globe-full of gas. The specific gravity is obtained from these weights of equal volumes of gas and air.

Fig. 15.

II. — OF LIQUIDS.

How may the Specific Gravity of a Liquid be found? — To obtain the specific gravity of a liquid, divide the weight of a convenient volume of it by the weight of an equal volume of water.

The experiments by which to find the weights of equal volumes of the two liquids may be made, —
1st, With a balance.
2d, With a hydrometer.
3d, With a solid bulb.

With a Balance. — The most direct method of getting the specific gravity of a liquid is to weigh equal quantities of it and water, and then divide the weight of the liquid by that of the water. To facilitate the operation, "specific gravity bottles" are made, which hold just one thousand grains of pure water. The weight of the bottle being known, a single operation with the balance will give the weight of the liquid, and then its specific gravity may be speedily calculated.

With a Hydrometer. — A common form of this instrument is represented in Fig. 16. It consists of a glass tube, with two bulbs near its lower end. The tube and upper bulb are full of air, which renders the instrument lighter than water. The lower and smaller bulb contains shot enough to keep the instrument in an erect position, when placed in a liquid, as shown in the figure. A graduated scale is fixed to the stem, to indicate the depth to which the instrument sinks in different liquids.

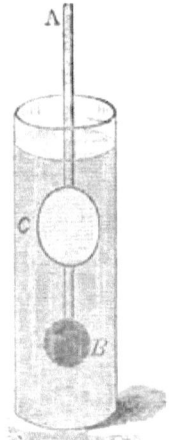

Fig. 16.

Explain the Action of this Instrument. — The action of the hydrometer can be readily explained by means of a piece of wood, several inches long and an inch square, having its lower end loaded with wire. If this be put into a vessel of water, it will sink to a certain depth, and remain upright. If it sinks ten inches, then ten cubic inches of water are displaced by it. If the instrument be put into a vessel of alcohol, it will sink deeper: suppose it be twelve inches; then twelve cubic inches of alcohol are displaced. But, according to the principle of Archimedes, the fluid displaced is equal in weight to the body floating in it.

Hence ten cubic inches of water have the same weight as twelve cubic inches of alcohol; or alcohol is $\frac{10}{12}$ as heavy as water. Its specific gravity is, therefore, $\frac{10}{12} = .833+$.

Making the instrument of glass, and giving it the form seen in Fig. 16, renders it more convenient and more accurate, but does not alter the principle on which it acts.

The Graduation.—The graduation of the scale is arbitrary, and varies in different forms of the instrument. The zero usually marks the point to which the hydrometer sinks in pure water, and the degrees above and below show how far the instrument may sink in liquids respectively lighter and heavier than water.

By the Use of a Bulb.—According to the principle of Archimedes, a heavy bulb of glass, or other convenient substance, when weighed in any liquid, will lose a part of its weight just equal to the weight of an equal bulk of that liquid. Hence, weigh a bulb of glass in air, afterward in water, and then in the liquid whose specific gravity is desired. The *losses* of weight it sustains will be the weights of equal bulks of the two liquids, and from these weights the specific gravity may be obtained.

To illustrate this method, suppose the specific gravity of alcohol is to be found. A bulb of glass, weighed in air and then in water, is found to lose 325 grains. Its loss in alcohol is found to be 257 grains. Then $\frac{257}{325} = .79+$ is the specific gravity of the alcohol. Alcohol, with this specific gravity contains no water: a higher specific gravity shows the presence of water.

III.—OF SOLIDS.

How may the Specific Gravity of a Solid be found?—To find the specific gravity of a solid body, divide the weight of a convenient portion of it by the weight of an equal volume of water.

There are two important cases of common occurrence:—

1st, The solid is heavier than water.
2d, The solid is lighter than water.

Of a Solid heavier than Water. — Divide the weight of the body in air, by its loss of weight in water. The principle of Archimedes explains this rule.

Thus the weight of a piece of marble in air is ten ounces, and in water it is 6.3 ounces: the difference, 3.7 ounces, is the weight of a bulk of water equal to the size of the marble. Then, $\frac{10}{3.7} = 2.7$ is the specific gravity of this solid.

Describe the Experiment. — The experiment is conducted in the following manner: Let the specific gravity of iron be desired. A fragment of iron of convenient size is hung from the bottom of one scale-pan of a balance, and weighed. It is then immersed in a vessel of water (see Fig. 17), and its weight again determined.

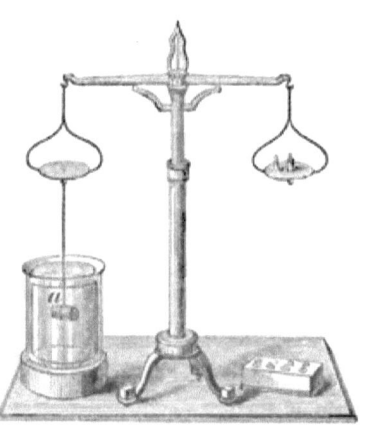

Fig. 17.

Of a Solid lighter than Water. — If the solid be lighter than water, the operation is more complex. If the light body be *compelled* to sink in water by fastening to it some heavier body, their loss of weight will represent the upward pressure of the water upon them both. If the heavy body alone be weighed in water, its loss will represent the upward pressure against it alone. If the upward pressure against the heavy body be subtracted from the upward pressure against both, the difference must represent the upward pressure against the light body alone, and hence the weight of a quantity of water equal to its bulk.

Illustration. — A body weighed 200 grains in air. When attached to a piece of lead, both weighed 1,936 grains in air, and 1,460 grains in water; suffering a loss of 476 grains. The lead itself, when weighed in water, lost 152 grains. The upward pressure against the light body alone must have been

$476 - 152 = 324$ grains. Then, 200 grains, the weight of the light body in air, divided by 324 grains, the weight of an equal bulk of water, is the specific gravity desired.

In the following table the specific gravities of several important substances are arranged for reference.

I.—Of Gases, at 32° F. Barometer, 30 inches.

Names.	Specific Gravity.	Names.	Specific Gravity.
Air	1.000	Nitrogen	0.972
Oxygen	1.106	Carbonic acid	1.529
Hydrogen	0.0692	Olefiant gas	0.978

II.—Of Liquids, at 39° F.

Names.	Specific Gravity.	Names.	Specific Gravity.
Water (distilled)	1.000	Ether	0.723
Sea-water	1.026	Naphtha	0.847
Milk	1.030	Oil turpentine	0.870
Alcohol (absolute)	0.792	Wine of Burgundy	0.991
Olive-oil	0.915	Mercury (32° F.)	13.598

III.—Of Solids, at 39° F.

Names.	Specific Gravity.	Names.	Specific Gravity.
Platinum (rolled)	22.069	Silver (cast)	10.47
Gold (stamped)	19.362	Diamond	3.50
Iron (cast)	7.2	Marble	2.837
Steel	7.8	Ivory	1.92
Lead (cast)	11.3	Flint-glass	3.329
Copper "	8.78	Ice	0.93
Brass	8.38	Pine wood	0.66

16. If an external pressure be exerted anywhere upon a liquid, the same amount of pressure will be transmitted in every direction.

It will act with the same force on equal surfaces, and in directions at right angles to them. This principle is known as PASCAL'S LAW. (*G.* 99, 109 ; *A.* 298, 299.)

The equal Transmission of Pressure. — To illustrate the principle stated above, let a vessel, represented in section by Fig. 18, be quite filled with water. In the sides of the vessel are several apertures, A, B, C, D, and E, closed with movable pistons. Let the area of each piston be one square inch, and suppose a weight of two pounds be placed upon the piston A. It will be found that the pressure on each piston *will be increased* by a force of just two pounds. Thus E will be pushed upward by a force of two pounds, while B and D will at the same time be pushed in opposite directions, each with a force of just two pounds. No matter how numerous these pistons may be, nor in what direction they may be inserted, each will be found exerting a two-pound pressure greater than before under the influence of a force of two pounds acting at A. *Every square inch in the entire surface of the vessel* will receive a pressure of two pounds.

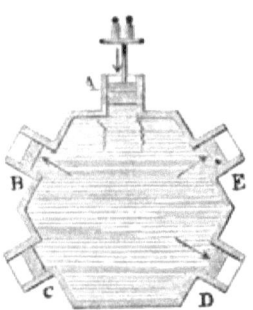
Fig. 18.

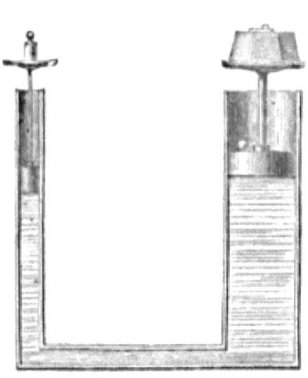
Fig. 19.

Experiment. — Then, suppose two cylinders, one just twice as large as the other, to be joined together by a tube at their bottoms (Fig. 19), and let there be a piston fitting each cylinder exactly, and carrying a table as shown in the picture. Now, according to the principle just learned, if a one-pound weight be put upon the small table, it will balance a two-pound weight upon the other.

If one cylinder be one hundred times larger than the other, then one pound on the small table will balance one hundred pounds on the large one.

The experiment shows that *if an external pressure be made on any part of the surface of a fluid, the pressure received by any other part of the surface will be in proportion to its area.*

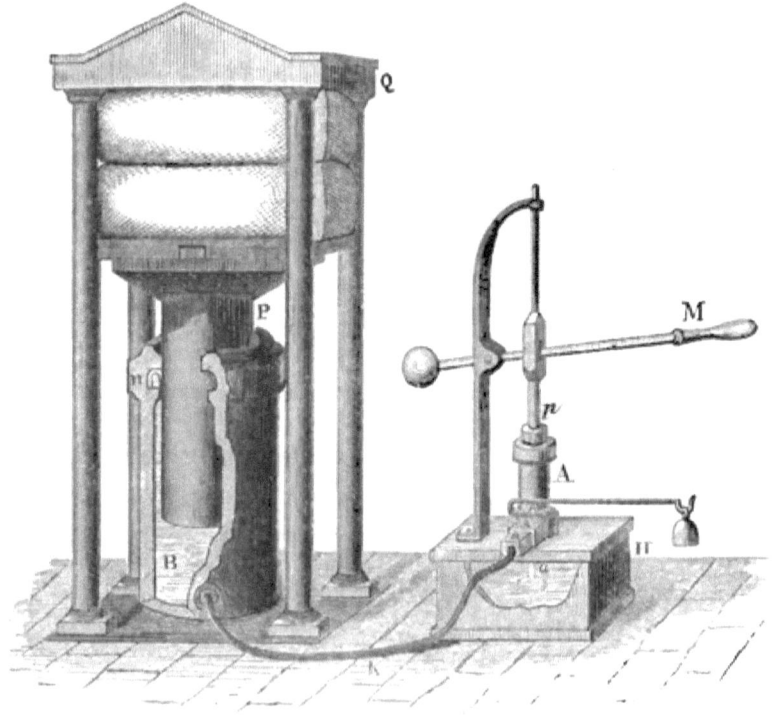

Fig. 20.

Application of this Principle. — The hydrostatic press acts upon the principle just explained. It is a machine by which a small force may be made to exert a great pressure. Its construction may be understood by examining Fig. 20.

Two metallic cylinders, A and B, of different sizes, are

joined together by a tube K. In the small cylinder there is a piston, *p*, which can be moved up and down by the handle M. In the large cylinder there is also a piston, P, having at its upper end a large iron plate which moves freely up and down in a strong framework, Q. Between the iron plate and the top of this framework, the body to be pressed is placed.

When the small piston is raised, the cylinder A is filled with water drawn from the reservoir H, below; and, when it is pushed down, this water is forced into the large cylinder, through the pipe K. There is a valve in this tube which prevents the water from returning, so that each stroke of the small piston pushes an additional quantity of water into the larger cylinder. By this means the large piston is pushed up against the body to be pressed.

To calculate the Pressure exerted by the large piston, we must remember that the force acting upon the piston in A will be exerted upon every equal amount of surface in B. To illustrate this, suppose the area of the large piston to be ten times the area of the small one; then one pound at A will produce a pressure of ten pounds at P.

By increasing the size of the large cylinder, and diminishing the size of the small one, the pressure exerted by a given power is increased proportionally. The weight of a man's hand may thus be made to lift a ship with all its cargo. The only limit to the increase of power is the strength of the material of which the machine is made.

SECTION V.

ON THE PROPERTIES OF GASEOUS BODIES.

17. The most characteristic properties of gases are Compressibility and Expansibility. Besides these, gases possess other properties common to all forms of matter, among which we notice Elasticity, Weight, and Mobility.

Compressibility. — The compressibility of air is prettily shown by the apparatus represented in Fig. 21. It consists of a glass tube with a bulb at the upper end, and with its lower end joined to another glass tube by a piece of rubber tubing. A colored liquid fills the bend, and stands at equal heights in the glass tubes. Closing the upper end of the open tube with the lips, let the breath be gently blown into it; the fluid will be seen to rise in the other tube, it may be a distance of several inches.

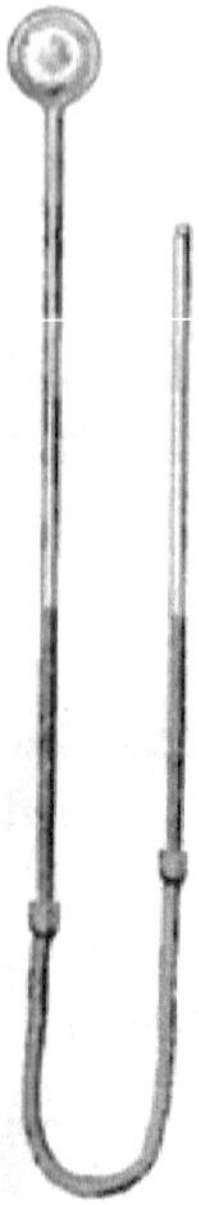

The air in the bulb can not escape, and the motion of the liquid shows that the air is being crowded into a smaller space; in other words, that it is *compressible*.

Expansibility. — If the bulb (Fig. 21) be warmed by grasping it in the hand, the colored liquid will move downward in the tube. The air in the bulb expands, and pushes the liquid along.

Fig. 21.

Or if, through the cork of a small bottle, a glass tube be passed, at the upper end of which is a bulb, and the lower end of which reaches down into the colored water contained in the bottle, the heat of a lamp-flame may be applied to the bulb. It will be noticed that bubbles of air escape from the lower end of the tube. The air is *expanded* so that the bulb and tube can no longer hold it all.

When the flame is withdrawn, the bulb gradually cools, and the water rises in the tube, and stands at a certain height, as shown in **Fig. 22**. Now let the palm of the hand be laid upon the bulb; the water is driven down the tube by the expanding air. The gentle warmth of the hand is quite sufficient to produce a very considerable expansion of the air.

These Properties are characteristic. — Solids and liquids likewise possess the properties of compressibility and expansibility in various slight **degrees**. All gaseous bodies possess them in very high degree. There seems to be no limit to the expansibility of gases, and the limit of compressibility is reached only when the gas is reduced to the liquid state.

By combined pressure and cold, all gases have been changed into liquids.

Elasticity. — The elasticity of air is beautifully shown by the simple apparatus already used (see Fig. 21) to illustrate the characteristic properties. Let the breath be alternately pressed into and withdrawn from the tube, the air will alternately be compressed, and spring back, which will be shown by the alternate motion up and down of the liquid in the tube.

Weight. — The air has weight. If we would show it, we may first weigh a closed vessel, properly arranged, and afterward take the air out of it, and weigh it again; the difference in these two weights will be the weight of the air which the vessel contains.

Fig. 22.

But how can the Air be taken out of a Vessel? — To answer this question, we must become acquainted with the *air-pump*. A section of the essential parts of this important instrument is represented in Fig. 23. A cylinder, A B, is joined by means of a tube, $b\ e$, to a very smooth plate, p. A piston, c, moves air-tight in the cylinder. In the piston is a valve, i, which opens upward, and another valve at b also opens upward from the tube into the cylinder. The vessel, d, from which the air is to be taken, is placed upon the plate. Such vessels are usually called receivers.

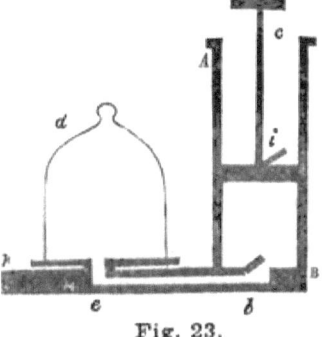

Fig. 23.

It will be seen that when the piston is raised, the valve, i, will be closed, and the air above it will be lifted out of the top of the cylinder. A vacuum would thus be formed below the piston, were it not for the expansibility of the air in the receiver. This air expands, and a part of it is forced through the valve, b, into the cylinder. When the piston is pushed down, the air below it passes through the valve, i; and when by a second stroke the piston is lifted, this air is pushed out at the top of the cylinder, while another portion from the

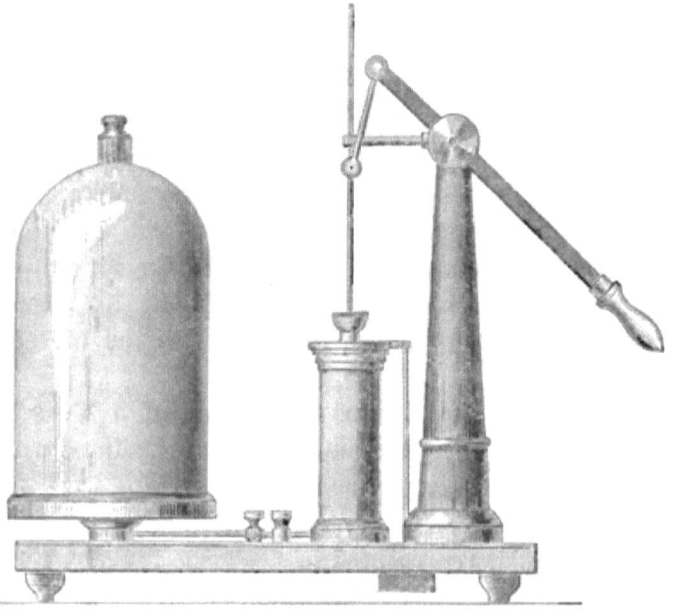

Fig. 24.

receiver is pressed through the tube into the cylinder below the piston. By each successive stroke, the quantity of air in the receiver is diminished, until, with a good instrument, the quantity left will be almost inappreciable. It is quite evident, however, that a *perfect* vacuum can not be obtained in this way. One form of this important instrument, complete, is represented in Fig. 24.

To weigh Air. — We may now attend to the process of weighing air. A hollow glass globe, with a stop-cock, is

NATURAL PHILOSOPHY. 47

hung from one pan of a delicate balance, and its weight carefully found. It is then screwed to the opening in the plate of the air-pump, and the air is exhausted. The stopcock is then closed to prevent the air from returning into the globe, which is then taken from the pump, and weighed. It

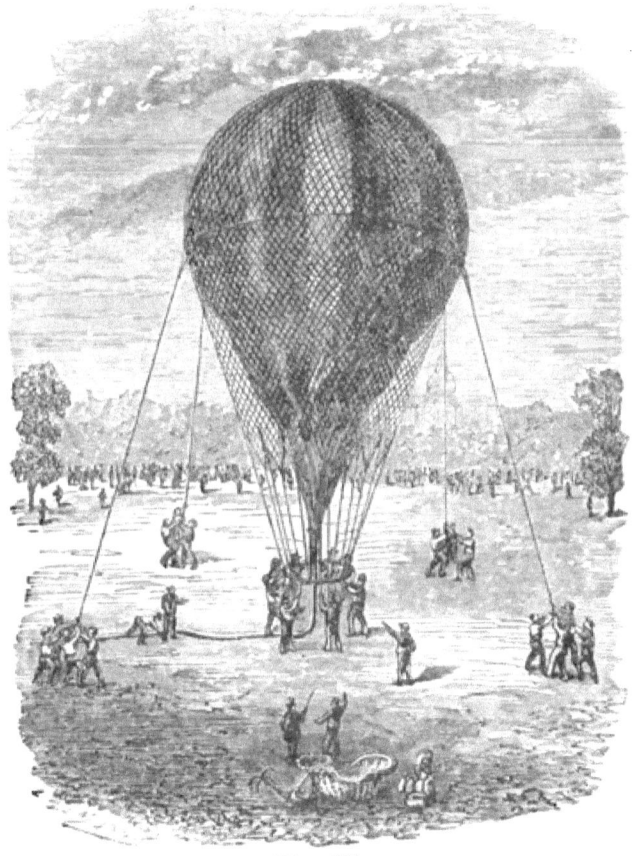

Fig. 25.

is found to weigh less than before, and the difference must be the weight of the air which has been taken out. At the ordinary temperature of air, one hundred cubic inches weigh about thirty-one grains.

Mobility. — The molecules of air and other gases move among themselves with the most perfect freedom. Their mobility exceeds that of liquids, because there is no cohesion at all to be overcome in them.

Pascal's law and the principle of Archimedes are therefore as applicable to gases as they are to liquids.

The Ascent of a Balloon is an exhibition of the principle of Archimedes. The balloon is filled with coal-gas or hydrogen, and it is lighter than an equal volume of air. It is therefore lifted by a pressure equal to the difference between the weight of the balloon with its contents, and that of an equal volume of air.

Absolute Weight. — Bodies show less than their real weight when weighed in air. They lose an amount equal to the weight of an equal volume of air. To obtain their true weights, bodies must be weighed *in vacuo*.

SECTION VI.

ON ATMOSPHERIC PRESSURE.

18. The atmosphere exerts pressure in all directions. This pressure is about fifteen pounds upon every square inch of surface. (*G.* 156, 157; *A.* 423, 424, 426).

The Atmosphere exerts Pressure. — Since every one hundred cubic inches of air weigh about thirty-one grains, it is clear that the atmosphere must be exerting considerable pressure upon the surfaces of all bodies on which it rests.

This Pressure may be shown. — Take a glass tube of convenient length, open at both ends, and insert one end in a vessel of colored water. Apply the lips to the other end, and, as the air is drawn out at the top, the water will be seen to rise rapidly in the tube. What pushes the water up? The ancients called it "Nature's abhorrence of a vacuum:" many at the present day are content to say that it is "sucked up." But let it be remembered that matter never moves unless it is *forced to move*, and that the forces of abhorrence and suction are simply fictions. The only force acting upon the water is the weight of the air resting upon its surface in the vessel. This downward pressure pushes the water under the lower end and upward into the tube.

It may be shown in another Way. — A more beautiful experiment consists in causing the pressure of the air to pro-

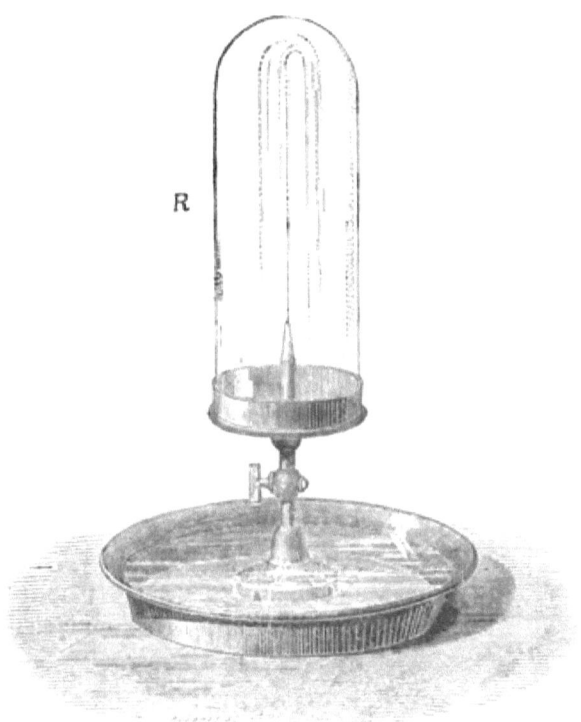

Fig. 26.

duce a fountain playing in a vacuum. A tall glass receiver (Fig. 26), closed at the bottom, has a stop-cock from which a tube extends upward a little way into the interior. This receiver is first exhausted by an air-pump, then fixed upon its base, and placed standing in a vessel of water; the stop-cock is opened, when instantly the water leaps to the top of the receiver, and a beautiful fountain continues to play until the jet-pipe is covered by the falling water.

The Pressure is in all Directions. — An experiment easily tried will show that the air is pressing equally in all directions. Stretch a piece of caoutchouc, or thin India-rubber, over the large end of a lamp-chimney, and firmly fasten it by winding a cord around it. Apply the mouth to the other end of the tube, and draw the air out. The pressure of the air pushes the rubber into the tube. Hold the tube so that the caoutchouc shall be above, or below, or sidewise in any direction, and *in all directions* alike the rubber will be pushed into the tube.

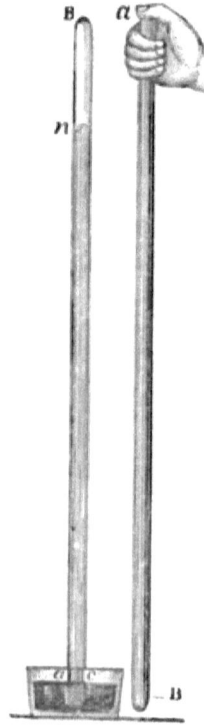

Fig. 27.

The Pressure is fifteen Pounds to the square Inch. — If the air should be all taken out of our tubes used in the foregoing experiments, to show that the atmosphere exerts pressure, the water would entirely fill them; and it is clear that the pressure of the atmosphere must at least equal the weight of the water which it lifts into the tube. How much farther the water would rise if the tube were long enough, these experiments have not told.

A *heavier* liquid will not be lifted as high as water, and will be more convenient for experiment. Mercury is a liquid metal about thirteen and a half times as heavy as water, and it is found that the air will sustain a column of mercury about thirty inches high. The experiment is conducted as follows: Take a glass tube more than thirty inches long, closed at one end, and fill it with mercury. Close the open end with the finger, and invert the tube. Next place the open end in a dish of mercury, and withdraw the finger. It will be seen that the top of the column of mercury in the tube is only about thirty inches above the surface of the mercury in the dish. (See Fig. 27.) The space above the mercury in the tube must be a vacuum. The experiment was first made by Torricelli, and this vacuum is called the TORRICELLIAN VACUUM.

Now, the pressure of the atmosphere just balances the weight of this column of mercury. But the weight of a column of mercury thirty inches high, the area of its base being one square inch, is fifteen pounds. The downward pressure of the atmosphere is therefore fifteen pounds to the square inch of surface on which it rests.

A Unit of Pressure. — Any pressure of fifteen pounds to a square inch of surface is called a pressure of ONE ATMOSPHERE. Heavy pressures are measured by this unit. A pressure of "four atmospheres" is a pressure of $4 \times 15 = 60$ pounds to a square inch.

At a temperature of 140° C. below zero, oxygen gas under a pressure of 320 atmospheres becomes a colorless liquid. What is this pressure in pounds?

The Kinetic Theory of Gases. — It is believed that the distances between the molecules of a gas are very many times greater than the diameters of the molecules themselves. Within these spaces the little molecules are darting about in all directions; but each one moves in a straight line until it strikes a neighboring molecule, or the side of the vessel in which the gas is held. On striking another, each molecule bounds away in another straight line until its course is again changed by another collision.

Pressure is the Energy of these Blows. — It is easy to see, that, if this theory is true, the molecules of air must be continually pounding against the walls of the vessel which contains it; or, indeed, against every thing which it touches. Every little blow exerts a little pressure, and such a multitude of swift blows is a constant pressure. Indeed, we believe that the pressure of air is nothing more than the energy of these molecular blows.

The Radiometer. — In the radiometer of Crookes (Fig. 28) this molecular energy actually turns a little wheel. The radiometer is a small glass globe with a very light vane resting on a pivot in the center. Each one of the four arms of the vane carries a disk of aluminum which is white on one side and black on the other. The air in the globe is exhausted to a very perfect vacuum. Of course there are then so few molecules, that they are able to fly in straight lines much farther before their path is changed by collision with one another.

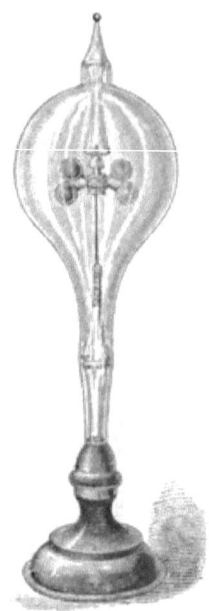

Fig. 28.

Now the sunbeam, or a candle-flame, will heat the black faces of the disks more than the white ones, and will make the vane whirl in a surprising manner, by thus increasing the energy of the molecular motion against the black faces.

19. The principle of atmospheric pressure is applied in the construction of many very useful instruments. We will notice the Barometer, the Common Pump, the Forcing Pump, and the Siphon.

I. — THE BAROMETER.

The barometer-column always indicates the pressure of the atmosphere. But the pressure of the atmosphere depends upon —

1st, Its height.

2d, The amount of water-vapor in it. (G. 162–167.)

The Barometer. — If the apparatus used to determine the pressure of the atmosphere (see Fig. 27) is inclosed for protection in a frame of metal or wood, with a graduated scale attached to measure the height of the column of mercury, it forms the instrument so well known as the BAROMETER.

Shows the Pressure of the Atmosphere. — The pressure of the atmosphere is not always the same. When it is less than fifteen pounds to the inch, the column of mercury will be lower than thirty inches, and, when greater, the column will be higher: indeed, the height of the column will vary in exact proportion to every change in the pressure of the air which supports it.

If the top of the mercury stands at 28 on the scale, the pressure of the atmosphere is $\frac{28}{30} \times 15$ pounds to the inch, nearly.

A Correction necessary. — We say *nearly* in this example, because the number on the scale at the top of the mercury column does not show exactly the true height. For we should notice that when the mercury sinks in the tube, it must rise in the cistern, so that the column must shorten at *both ends*. The figures on the scale, however, only show the change which takes place at the top: they fail to tell the true height of the column.

Made by Fortin's Cistern. — This error is avoided in what is called Fortin's barometer, by means of a cistern with a flexible bottom. (See Fig. 29.) The bottom of this cistern is made of deer-skin, and rests upon the end of the screw C, by which it may be lowered or lifted. An ivory pointer, A, is fastened to the top of the cistern, and its lower end is the point from which the distances are measured on the scale which shows the height of mercury in the tube. If the surface of the liquid in the cistern just touches this point, then the figures on the scale show the true height of the column, which indicates the pressure of the atmosphere.

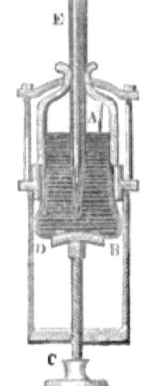

Fig. 29.

The Pressure of the Atmosphere depends upon its Height. — When we go up a mountainside, we leave a part of the atmosphere below us, and of course the height of the column above us is less. Hence the weight of the atmosphere will vary with the altitude of the place where the observation is made, and on this account the height of the barometer column will be different at different distances above the sea-level.

To Measure Heights of Mountains.— Upon this principle the barometer is *used to measure the heights of mountains.*

If the density of the atmosphere were uniform, the fall of the mercury would be in the exact ratio of the distances upward; and, knowing the height required to make the mercury fall one-tenth of an inch, this, multiplied by the number of tenths through which it is observed to sink, would tell the height of the mountain. The truth is, however, that the density of the air rapidly diminishes as we ascend. Temperature, too, affects its pressure. In spite of these difficulties, tables have been constructed by which the height of a place above the sea-level may be calculated by observing the height of the barometer-column and the temperature of the atmosphere.

Pressure depends also upon the Amount of Water-Vapor present.— Mixed with the air, at all times, are considerable quantities of *invisible* vapor of water. If the atmosphere were pure dry air alone, it would exert a certain pressure; if it consisted wholly of water-vapor, it would exert a different amount of pressure: it does consist of a mixture of these two gases, and the pressure it exerts is the *sum* of the pressures they would separately exert.

It follows that the atmospheric pressure will be greatest when there is the greatest quantity of water-vapor in the air: the barometer-column will then rise. But let this vapor be condensed into clouds, and it will have but little force of elasticity, and will exert but a small fraction of its former pressure: hence the barometer-column will stand lower in cloudy weather.

To predict Weather Changes.— On this principle the barometer is used to indicate changes in the weather. *A rising column indicates fair weather; a falling column indicates foul weather.*

This rule is to a great extent reliable. Others are added by different observers, but they must be taken with considerable allowance.

II.—THE COMMON PUMP.

Description.—This instrument, as generally made, consists of two cylinders or barrels, A and B, Fig. 30, with a valve, S, at their junction, opening upward. In the upper barrel is a piston, P, in which is a valve, O, also opening upward. The piston is moved by means of the handle H, and the water may flow from the spout C.

Explanation.—When the piston is lifted, the air above it will be lifted out of the barrel. A partial vacuum will thus be formed below the piston, and the pressure of the air upon the surface of the water in the well will push the water up the barrel A, through the valve S, into the barrel B. When the piston goes down, the valve S will close, and prevent the return of the water to the well. The valve in the piston will be opened, and the water will pass through it. When the piston is again lifted, the water now above it will be lifted to the spout, while the atmospheric pressure will force another portion into the barrel below the piston.

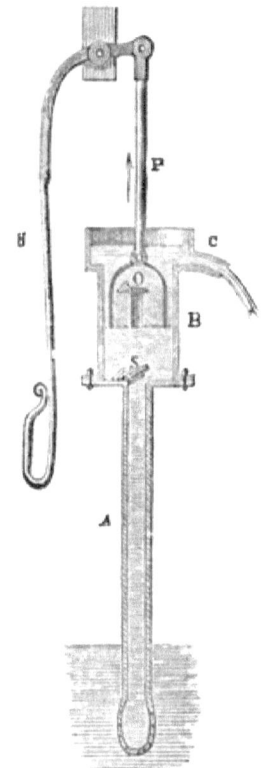

Fig. 30.

Limit of Height.—At the sea-level the pressure of the air will sustain a column of mercury about thirty inches high. Since mercury is, at ordinary temperature, about $13\frac{1}{2}$ times heavier than water, the same force will lift a column of water $13\frac{1}{2}$ times as high: $13\frac{1}{2} \times 30 = 405$; 405 in. $= 33\frac{3}{4}$ feet. The atmosphere can not lift water in the common pump to a greater height than this, even at the level of the sea. In practice the valve S must be less than $33\frac{3}{4}$ feet above the water.

III.—THE FORCING PUMP.

Description.—In the forcing pump the piston has no valve, but from near the bottom of the upper barrel there is a tube passing to an air-chamber, with a valve opening into the chamber. A section of this instrument is represented in Fig. 31. Reaching from near the bottom of the air-chamber, I K, is a tube, L M, which extends to any place at which the water is to be delivered.

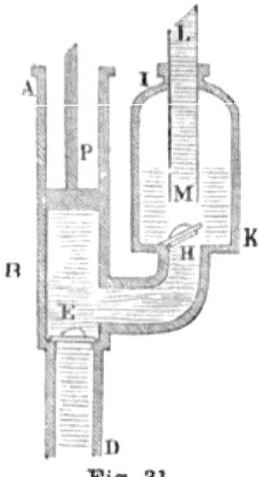

Fig. 31.

Explanation.—Now, when the solid piston P is raised, water is pressed through the valve E, into the barrel B. When the piston is pushed down again, the water is driven through the tube into the air-chamber, and compresses the air in it. By every stroke, the water accumulates in the chamber, and the air is more and more compressed. The pressure of this condensed air upon the water in the chamber pushes it up through the tube L M, to the place where it is desired. Without the air-chamber, the water would issue from the pipe in jets: with the chamber, the water issues in a steady stream.

IV.—THE SIPHON.

Description.—The siphon is an instrument by which liquids may be transferred from one vessel to another, by atmospheric pressure. It consists of a bent tube, one arm of which is longer than the other. In Fig. 32 the siphon in operation is shown. Having been first filled with water, its short arm is inserted in the water to be transferred from the vessel C, and it is then found that the water will flow steadily until the lower end of the short arm is left uncovered, or, in other cases, until the water in the two vessels stands at the same level.

Explanation. — The downward pressure of the air at C is partly balanced by the weight of the column of water, C D, in the short arm of the tube; the *excess* of pressure will tend to push the water over through the bend toward B. On the other hand, the atmospheric pressure at B is partly balanced by the pressure of the water in the long arm, which is equal to the weight of a column A B; the *excess* will tend to push the water back through the bend toward C. It is clear that the pressure of air, minus the weight of the *shorter* column of water, is more than the same pressure, minus the weight of the *longer* column, and hence that a greater force will be exerted to push the water from C toward B, than from B toward C: the liquid will flow in the direction of the greater force, up the short arm, over the bend, M, and out at B.

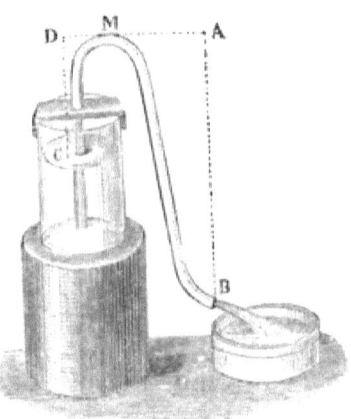

Fig. 32.

SECTION VII.

ON "THE THREE LAWS."

There are three fundamental principles which apply to all true gases. They are known as

The Law of Boyle or Marriotte,
The Law of Charles,
The Law of Avogadro.

I. — THE LAW OF BOYLE.

20. The volume of a given weight of air will be inversely as the pressure upon it. (*G.* 174, 175; *A.* 412, 458.)

Volume depends upon the Pressure. — Press the breath into the open tube of the apparatus represented in Fig. 21, and the liquid flows toward the bulb, showing that the air in the bulb is *condensed*. Next draw the air out of the tube, and the liquid flows away from the bulb, showing that the air within is *expanded*. The same quantity of air is here seen to fill *less* space when the pressure upon it is *increased*, and *more* space when the pressure is *diminished*.

Plan for further Experiment. — Now, we may prove by experiment, first, that with a double pressure, the volume will be just one-half; and, second, that with half the pressure, the volume will be just double.

Pressure greater than one Atmosphere. — In the first case, we use a bent glass tube (Fig. 33), the short arm being closed, and the other, which should be more than thirty inches long, being open at the top. A graduated scale, to which the tube is firmly bound, measures inches from the bend.

Now let mercury be poured into the tube, until it fills the bend. The air presses upon the mercury in the long arm, and this liquid transmits the same pressure to the air in the short arm. The pressure upon the air in the short arm is, therefore, fifteen pounds to the square inch. If we fill the long arm with mercury, as shown in the figure, to the height of thirty inches above the level of the mercury in the other, we shall be adding a pressure of fifteen pounds to the inch.

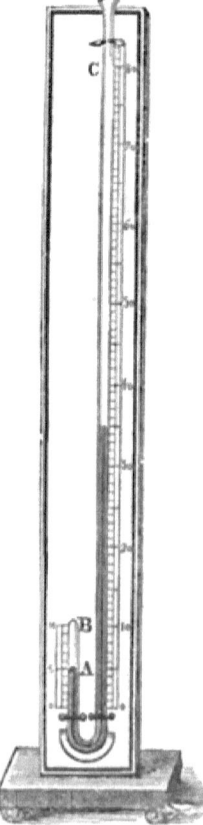

Fig. 33.

The pressure upon the air in the short arm will then be *doubled;* and we shall discover that the mercury has risen, crowding the air before it, and stands at A, the air having just *half* its original volume.

Pressures less than one Atmosphere. — In the second case, we take a glass tube (A B, Fig. 34), about twenty-five inches long, and open at both ends. Let three narrow bands of paper be pasted upon it, the first at a distance of three inches from the top, the second six inches from the top, and the third fifteen inches from the second. Let another larger tube, D, about thirty inches long, be nearly filled with mercury. Insert the end, A, of the small tube, in this mercury, and push it down until the upper mark (3) is at the level of the mercury. Now, clasping the finger tightly over the end B, thus inclosing three inches of air in the tube, lift it until the third mark is brought up to the top of the mercury. The air will be found to fill the space of six inches.

Before the tube was lifted, the whole pressure of the atmosphere, fifteen pounds to the inch, was exerted upon the air within: after the tube was lifted, the atmosphere sustained a column of mercury fifteen inches high. To do this, required half the pressure it can exert: the other half was exerted upon the six inches of air above the mercury. We thus show that with *half* the pressure the volume will be just *double*.

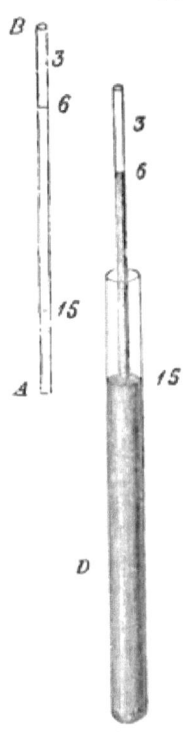

Fig. 34.

What is true in both Cases? — In these two experiments we find that *the volume of a given weight of air is inversely as the pressure upon it;* and repeated experiment confirms the inference that the same principle holds true for other pressures.

This law was discovered independently by Robert Boyle in England, and the Abbé Marriotte in France.

The Density of the Atmosphere. — When a given weight of air is crowded into one-half its original volume, it must be twice as dense; and, when expanded into double its first volume, it can only be half as dense. The density of air will therefore be exactly in proportion to the pressure upon it. So the atmosphere, where its own pressure is greatest, will be most dense.

Is greatest at the Surface of the Earth. — The atmosphere in contact with the earth is pressed upon by all the air above, even to the top of the atmosphere. At a distance above the earth, the atmosphere receives less pressure, because there is less air above to exert it. The density being greatest where the pressure is greatest, the air at the surface must be more dense than the portions above. The air is much less dense at the top of a high mountain than at its base.

II. — THE LAW OF CHARLES.

21. The volume of a given weight of air will be greater as its temperature is higher. It expands $\frac{1}{491}$ of its bulk at 32° for every additional degree Fahrenheit, or $\frac{1}{273}$ of its bulk at 0° for each degree Centigrade.

Charles's law declares that the volume of a given weight of gas is directly as its absolute temperature.

Heat increases the Volume of Air. — Let the palm of the hand be laid upon the bulb (Fig. 22), and the fluid in the tube descends, because the air in the bulb expands. Pour cold water upon the bulb, and the fluid ascends, because the air above it is condensed. Apply the heat of the lamp-flame to the bulb, and the water in the tube will be quite driven out at the bottom: let it cool again, and the water rises to its former height. These experiments show that the addition of heat expands air, and that its withdrawal contracts it.

At the Rate of $\frac{1}{491}$ its Bulk for each Degree. — The expansion of air and other gases, by heat, is uniform. One degree of heat when the temperature is low produces the same expansion as one degree when the temperature is high. If we have 491 cubic inches at a temperature of 32°, it will become 492 cubic inches if heated one degree, making its temperature 33°. In other words, it expands $\frac{1}{491}$ *of the volume at* 32°, for each additional degree of heat applied.

If the Centigrade scale of temperature be used, then the illustration is as follows: 273 cubic inches of gas at 0° will become 274 cubic inches if heated to 1°. That is to say, it expands $\frac{1}{273}$ of *its volume at* 0° for each additional degree.

One hundred cubic inches of air at 32° F. will be expanded at 60° F. to be $100 + \frac{28}{491} \times 100 = 105.7$ cubic inches.

The Effect of Cooling. — Conversely each degree of heat taken away from a gas, beginning at 0° C. or 32° F., will *reduce* its volume at the same rates. 273 cubic inches at 0° C. will be 272 cubic inches at $-1°$ C., and 271 cubic inches at $-2°$ C. The gas would go on losing $\frac{1}{273}$ of its volume for each degree of lower temperature. Then

How many Degrees can be withdrawn? — Not *more* than 273, because it would be impossible for the gas to contract *more* than $\frac{273}{273}$ of its volume. If the law holds good at such low temperature, then at $-273°$ C. the volume of a gas would be infinitely small, and it would be impossible to conceive it any colder.

Absolute Temperature. — This point on the Centigrade scale, $-273°$, is called the *absolute zero*, and temperatures measured from this point are called the *absolute temperatures.*

Absolute temperatures are obtained by adding 273 to the temperature on the Centigrade scale. Thus 60° is 333° on the absolute scale, and $-40°$ is 233.°

The Law of Charles declares that *the volume of a given weight of gas, the pressure being unchanged, will vary directly as the absolute temperature.*

III. — THE LAW OF AVOGADRO.

22. Equal volumes of all gases, at the same temperature and pressure, contain the same number of molecules.

Illustration. — It is believed that there are just as many molecules in a cubic inch of oxygen as there are in a cubic inch of hydrogen, when these gases are under the same pressure and at the same temperature.

The proof of this law is found in the study of mathematical physics and of the science of chemistry. We must therefore omit the demonstration now; but the law itself should be remembered with those of Boyle and Charles, because these three laws together define the truly gaseous condition of matter.

SECTION VIII.

REVIEW.

I.—SUMMARY OF PRINCIPLES.

The solid, liquid, and gaseous conditions depend on the relative strength of attraction and repulsion between the molecules of the body. Hence,

Hardness, tenacity, malleability, ductility, and crystalline form are properties of solids only; while

Mobility is the characteristic property of liquids, and

Compressibility and expansibility are the characteristic properties of gases.

The pressure of a liquid at rest is due to gravitation, and is exerted equally in all directions. For this reason,

The surface of water at rest must be level. The liquid will always rise to the same height in all tubes or reservoirs which communicate with one another.

The pressure of a liquid on the base of the vessel containing it depends upon two things *only*,—the area of the base, and the height of the liquid above the base. It is equal to the *weight* of a column as *large as the base* of the vessel and as *high as the surface* of the liquid.

A body immersed in fluid, either liquid or gaseous, is pressed both upward and downward; but the upward pressure is the greater. Hence a body loses weight when immersed, and its loss is just equal to the weight of the fluid it displaces.

NATURAL PHILOSOPHY. 63

A fluid, either liquid or gaseous, transmits external pressure in every direction, and allows it to act with the same energy on every equal amount of surface.

II.— SUMMARY OF TOPICS.

8. The molecular forces. — Illustrated in natural changes in condition. — In artificial changes.

9. Hardness. — Tenacity. — Malleability. — Ductility. — Crystalline form. — These properties characteristic of solids.

10. Elasticity of liquids. — Attraction and repulsion nearly equal. — Mobility. — Characteristic property.

11. Liquids press in all directions. — Equally. — The surface at rest must be level. — Level surface convex.

12. Water in pipes rises as high as its source. — The supply of water to cities. — Springs. — Artesian wells.

13. Pressure independent of the shape of the vessel. — It depends on the depth of the liquid. — On the area of the base. — To calculate the pressure. — Against any surface.

14. Bodies immersed are pressed upward. — With force equal to the weight of fluid displaced. — The solid lighter than water. — The solid heavier than water.

15. Specific gravity. — Of gases. — Of liquids. — By direct weighing. — By the hydrometer. — By the use of a bulb. — Of solids. — Heavier than water. — Describe the experiment. — Lighter than water. — Table.

16. The equal transmission of pressure. — Experiment. — The shape of the vessel makes no difference. — The hydrostatic press.

17. Of the properties of gases. — Compressibility. — Expansibility. — These properties characteristic. — Elasticity. — Weight. — The air-pump. — To weigh air. — Mobility. — Ascent of a balloon. — Absolute weight.

18. The atmosphere exerts pressure. — Shown by experiment. — In all directions. — About fifteen pounds to the square inch. — Unit of pressure. — The Kinetic theory. — The radiometer.

19. The barometer. — Shows the pressure of the atmosphere. — Correction needed. — Made by Fortin's cistern. — The pressure of the atmosphere depends upon its height. — Measurement of mountains. — Upon the amount of water-vapor it contains. — To predict changes in the weather. — The common pump. — The forcing-pump. — The siphon.

20. Volume of gas depends on pressure. — Plan for experiment. — Pressure greater than one atmosphere. — Pressure less than an atmosphere. — Law deduced. — Density of the atmosphere. — Density greatest at surface of the earth.

21. Heat increases the volume of air. — At the rate of $\frac{1}{491}$ of its bulk for each degree. — The effect of cooling. — How many degrees can be withdrawn? — Absolute temperature. — Law of Charles.

22. Law of Avogadro. — Illustration.

III. — PROBLEMS.

Problems illustrating the Laws of Hydrostatics.

1. A cylindrical vessel, whose base is 5 square feet, is 10 feet high. It is filled with water. What pressure is exerted upon the base? *Ans.* 3125 pounds.

2. If the bottom of a vessel has an area of 72 square inches, and its top an area of 96 square inches, and it is 9 inches high, what pressure will be exerted on the bottom when the vessel is full of water? *Ans.* 23.43+ pounds.

3. Two vessels with equal bases are filled with water, one to a height of 9 inches, the other to a height of 27 inches. How many times more pressure on the base in the last case than in the first? *Ans.* 3.

4. How much pressure is exerted against the *side* of a cubical vessel which is full of water, its height being 18 inches? *Ans.* 105.46+ pounds.

5. How much pressure would be exerted upon 12 square inches of the sides of the vessel when the middle point of this surface is 20 inches below the top of the water?

Ans. 8.68 pounds.

NATURAL PHILOSOPHY. 65

6. How many cubic inches of water will be displaced by a piece of pine wood weighing just 10 pounds?
Ans. 276.48.

7. How much less will a piece of marble measuring 100 cubic inches weigh in water than in air? *Ans.* 3.61 pounds.

8. The specific gravity of marble being 2.7, what will be the weight of 25 cubic feet? *Ans.* 4218.75 pounds.

9. How many cubic inches in a block of ice that weighs 75 pounds? *Ans.* 2229.67.

10. What is the specific gravity of flint-glass if a fragment of it weigh, in air, 4320 grains, and in water 3023 grains? *Ans.* 3.33.

11. The specific gravity of wax is to be found from the following data: —

Weight of the wax in air . . . . 8 ounces.
Weight of a piece of lead in air . . 16 ounces.
Weight of the lead in water . . . . 14.6 ounces.
Weight of wax and lead in water . 13.712 ounces.
Ans. 0.9.

12. A bottle holding 1000 grains of water is found to hold only 870 grains of oil of turpentine. What is the specific gravity of this oil? *Ans.* 0.87.

13. How much pressure can be exerted upon the large piston of a hydrostatic press by applying fifty pounds to the small piston; the area of the small piston being one-half square inch, that of the large piston 100 square inches?
Ans. 10000 lbs.

Problems illustrating the Laws of Gaseous Bodies.

1. What is the weight of a cubic foot of air at ordinary temperature and pressure? *Ans.* 535.68 grains.

2. What is the weight of 100 cubic inches of oxygen gas at ordinary temperature and pressure, its specific gravity being 1.106? *Ans.* 34.286 grains.

3. What is the weight of 100 cubic inches of nitrogen gas at ordinary temperature and pressure, its specific gravity being .972?

4. What pressure will be exerted by the atmosphere on a surface of one square foot? *Ans.* 2,160 pounds.

5. What pressure does the atmosphere exert upon a square inch surface when the barometer column is 28 inches high?
Ans. 14 pounds.

6. How high a column of water would the atmosphere sustain when the barometer column stands at a height of 28 inches? *Ans.* $31\frac{1}{2}$ feet.

7. Suppose 100 cubic inches of air at a pressure of 15 pounds to the inch is made to receive an additional pressure of 15 pounds to the inch: what will be its volume?
Ans. 50 cubic inches.

8. How much pressure must be removed from 100 cubic inches of air, at usual density, in order that it may expand to a volume of 200 cubic inches?
Ans. $7\frac{1}{2}$ pounds to the inch.

9. In the air-chamber of the forcing-pump the air is compressed into half its former bulk; how high will the water be thrown? *Ans.* $33\frac{3}{4}$ feet.

10. If we have 500 cubic inches of air at 32° F. temperature, how much will there be when it is heated to a temperature of 75°? *Ans.* 543.78+ cubic inches.

CHAPTER III.

ON MOTION.

23. The Fundamental Ideas. — ATTRACTION and REPULSION acting upon masses or MOLECULES of matter determine their condition of rest or motion.

Illustration. — The motion of bodies falling to the ground is due to the *attraction* of gravitation. The motion of air in wind is caused chiefly by the *repulsive* power of heat. The bullet speeds on its way, urged by the *repulsive* force of exploding gunpowder. The forces which produce the endless variety of motions in nature are found, when carefully studied, to be only different forms of attraction and repulsion.

SECTION I.

ON MOTION CAUSED BY A SINGLE FORCE.

24. There are three important principles known as Newton's laws of motion: —

1st, A body at rest will remain at rest; or, if in motion, it will move forever in a straight line, unless acted upon by some force to change its condition.

2d, A given force will produce the same amount of motion, whether it act upon a body at rest or in motion.

3d, Action and reaction are equal, and in opposite directions. (*A.* 154-157, 168.)

The first Law. — The truth of the first law is seen when we remember that the inertia of matter forbids that a body shall, in any way, change its own condition.

Then, why are bodies so constantly changing their condition of rest or motion? We rarely see a body in nature, moving in an absolutely straight line? The reason is this: bodies are constantly under the influence of *forces* which do change their condition. A stone thrown from the hand would move forever in a straight line, if it felt only the force of the hand; but gravitation, and the resistance of the air, compel it to move in a graceful curve instead. The pleasing variety of natural motions is brought about by the unceasing action of external forces.

The second Law. — Let us examine this law by means of the diagram, Fig. 35. If a ball be thrown suddenly from the point A, horizontally, it would go to the point B, if it could be let alone by other forces. So likewise, if it be dropped from A, gravitation alone will carry it to C. Now, these separate effects of the two forces will be exactly produced when the forces act together. Suppose the ball to go from A to B in one minute, and, when dropped, to fall from A to C in the same time. While the ball is moving toward B, gravitation is pulling it downward, and at the end of the minute it will be found at D, having moved to the right a distance exactly equal to A B, and downward through a distance exactly equal to A C; so that the force of gravitation produces the same effect, whether it act upon the ball *resting* at A or *in motion* toward B.

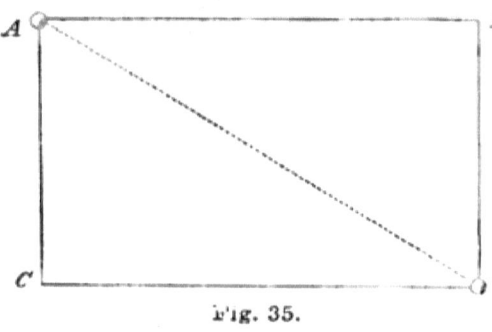

Fig. 35.

The third Law. — If a table be struck, the hand that strikes it receives a blow as well. The hand *acts* upon the table; the table *reacts* upon the hand. Attend, now, to the following experiment. Two ivory balls are suspended by cords, and hang in contact against a graduated arc. When the ball B is lifted up the arc to D, and then allowed to swing against the other, it strikes it, and instantly stops, while the other ball takes up its motion, and goes to the point C. The first ball acts upon the second; the second reacts upon the first. Now, if we notice that the motion from D to B, which is *stopped* by the reaction of the second ball, is just equal to the motion from A to C, which is *caused* by the action of the first, it becomes evident that the two forces must be equal, and exerted in opposite directions.

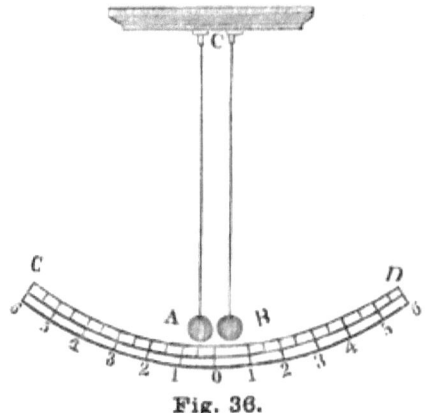

Fig. 36.

It follows from this principle, that, when two bodies come in contact, each one gives and receives an equal shock. The hand which strikes the table is itself bruised, and the bullet which shatters the bone is itself battered.

25. The velocity of a moving body will be uniform if it be produced by an impulsive force, and opposed by no resistances.

The elements of motion are Time, Space, and Velocity. In uniform motion, the space is equal to the product of time multiplied by velocity. (*G*. 25; *A*. 105, 110.)

Velocity. — Velocity, in a popular sense, is simply rapidity of motion; but if the term is to be of any scientific value it must be more definitely applied. Velocity is the *distance* passed over by a body in a *unit of time*. The velocity of a cannon-ball, for example, may be two thousand feet a second; that of a train of cars may be thirty miles an hour.

Uniform Velocity. — In uniform velocity, a body moves

over equal spaces in equal times. If, for instance, in each of three successive hours, a steamboat travels fifteen miles, its velocity is uniform.

An Impulsive Force. — An impulsive force is one which, after acting for a time, ceases. The stroke of a bat, which knocks the ball, is an impulsive force; so are the blows of a hammer. No matter how long a force may have been acting, if it be suddenly withdrawn, it is at that moment an impulsive force.

Uniform Motion produced by an Impulse. — If a body can be free from all forces but the impulse which gives it motion, its velocity will be uniform. This seldom occurs. How rarely do we see a uniform motion produced by an impulse, either in nature or in art! This is because all bodies are under the influence of several forces at once, such as gravitation, friction, and the resistance of air, by which their velocities are changed. The motion of the earth on its axis is, however, a sublime example of uniform motion.

In the arts, a uniform motion can be secured only by the constant application of power. The impulse which starts a train of cars would make it move uniformly if it did not meet with resistances; to overcome these, a constant pressure of the steam must be applied. If this pressure be at all times just enough to accomplish that purpose, the motion of the train will be uniform.

Space equals Time multiplied by Velocity. — It is evident that a train of cars, going uniformly at the rate of 25 miles an hour, will in 10 hours go 250 miles. But we see that $250 = 25 \times 10$, or, that the space is equal to the product of the two other elements, time and velocity.

We may express this principle by the simple equation, —
$$S = T \times V:$$
in which S stands for Space; T stands for Time; V stands for Velocity.

Now, if any two of these elements are given, the third may be found by *substituting the given values for the letters,* and then *performing the operations indicated.* For example, what is the velocity of a bullet which flies 2,000 feet in 20 seconds, supposing its velocity uniform? The value of S is 2,000 feet; the value of T is 20 seconds. Putting these values in the equation, it becomes

$$2,000 = 20 \times V. \text{ Hence } V = 100.$$

26. The motion of a body produced by the action of a constant force alone, will be uniformly accelerated. The difficulties in the way of any accurate experiment upon uniformly accelerated motion are overcome by Atwood's machine. (*G.* 27, 77, 78; *A.* 132–135.)

A constant Force. — By a constant force we mean a force which acts upon a moving body all the time alike. The force of gravitation is the most perfect example of a constant force.

Uniformly accelerated Motion. — The motion of a body is uniformly accelerated, when its velocity increases equally in successive units of time, as, for example, five feet the first second, eight feet the next second, eleven feet the third second, fourteen feet the fourth, and so on.

The motion of a falling body is the most perfect example known of uniformly accelerated motion. It would be a perfect example, were it not for the resistance of the air.

Difficulties in the way of Experiment. — But there are three difficulties in the way of accurate experiments upon the motion of a body falling. 1st, This motion is so *rapid* that no accurate observations can be made. 2d, It is subject to the *resistance of air*, which reduces its velocity. 3d, The *friction* of any apparatus used is likely to impede it.

These Difficulties overcome by Atwood's Machine. — These difficulties are, for the most part, overcome by Atwood's machine (Fig. 37). Two heavy weights, A and B, are fastened to the ends of a small cord which passes over a grooved wheel, D. Each end of the axis of this wheel rests on the circumferences of two other wheels. This set

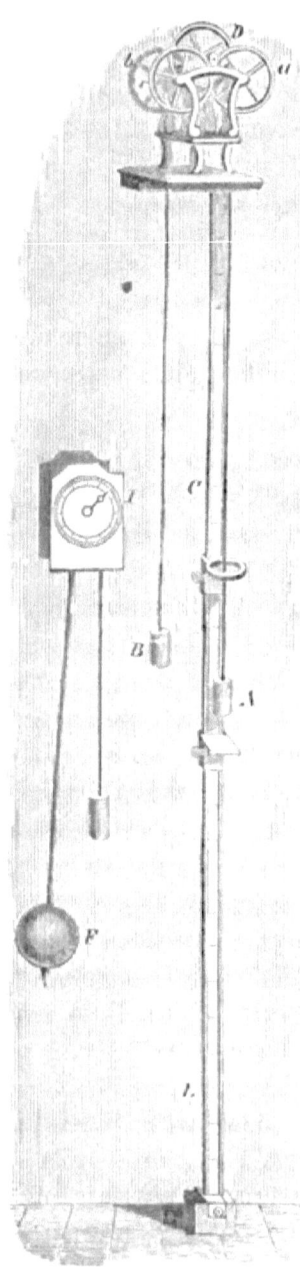

Fig. 37.

of wheels may be supported by a bracket firmly fixed to the wall of a room, several feet above the floor. The standard C L, reaching to the floor, is graduated; upon it is a movable ring, which allows the weight A to pass through it; and a table below, which arrests the motion of the weight at any desired point. The time of motion is measured by the pendulum F.

The two weights A and B are made exactly equal, and of course, when left to themselves, will remain at rest. But, if a small bar of brass be laid upon the weight A, motion takes place, due entirely to the action of gravitation upon the bar.

Now, suppose the large weights each to be $31\frac{1}{2}$ ounces, and the weight of the small bar to be one ounce. When they all move, 64 ounces are in motion, but this motion is caused by the force which acts upon the one-ounce bar. It is evident, that, if the force is the same, 64 ounces will move only $\frac{1}{64}$ as fast as one ounce. The motion of the weights is produced by a *constant force*, gravitation; but it is only $\frac{1}{64}$ as rapid as when the bodies fall freely. A *slow* motion is thus obtained. The *resistance of the air* against the small surfaces of the ends of the heavy weights is very slight when they move slowly. The *friction* of the wheels at the top is trifling. And thus the three difficulties in the way of experiment are overcome.

27. By experiments with Atwood's machine we may prove:—

1st. That a body moving under the influence of gravitation during any interval of time will gain a velocity which, acting alone, will carry the body twice as far in the next equal interval.

2d, That gravitation will add to the motion of a body just as much in every interval of time as it produced in the first.

By the help of these principles we may analyze the motion of a falling body. From the diagram which represents this analysis, we may construct a table which shall contain the values of Time, Space, and Velocity; and from this table obtain the Laws which govern the motion, and the Formulas by which problems may be solved.

Proof of the first Principle.—Let the weight A, carrying the small bar, be brought to the top of the graduated standard, and let the ring C be placed three inches below. Suppose that, in one second after its release, A falls to the ring. The small bar will be caught off by the ring; the weight A will pass through, and in the next second it will be found to go exactly six inches. By putting the ring at different places on the standard, it will be found that in every case, as in the one just described, the *body moving under the influence of gravitation during any interval of time will gain a velocity which, alone, will carry the body twice as far in the next equal interval.*

Proof of the second Principle.—If the weight and bar fall three inches in one second, they will be found by experiment to fall twelve inches in two seconds. Hence the distance fallen in the *second interval* is nine inches. If the bar were taken off at the end of the first second, we know that the weight would go alone six inches in the next. It is clear, then, that the bar acting in the last second adds a motion of three inches, the same amount as it produced in the first. Repeated experiments show that *gravitation will add to the motion of a falling body just as much in each second as it produced in the first.*

Analysis of the Motion of a falling Body.—Now suppose a body to fall from the point A (Fig. 38), toward the point D. In the first second it will fall a certain distance, which we will represent by A B. For a moment suppose the force of gravity should cease to act: the body would still move on, and we know (by the first principle) that it would go in the next second just twice as far as it did in the first. Then mark below B two spaces, each equal to A B, to represent this distance, and mark it with a heavy line, that the eye may see at a glance that it is the distance due to velocity alone. But we know (by the second principle) that gravitation in this second will add a space just equal to A B. Marking this additional space in the figure, we find that in two seconds the body will fall to C.

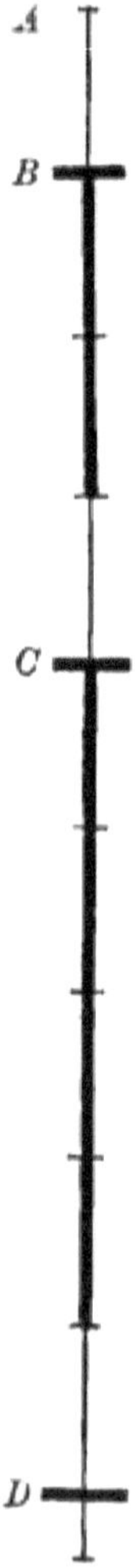

In *two seconds* the body has fallen four spaces: in the *next two seconds* it will go twice as far, eight spaces, by velocity alone. In the *first* of these two seconds, which is the *third second* of its fall, the body will go one half that distance, or four spaces, by velocity. The force of gravity adds another space, so that at the end of three seconds the body will be found at D.

To find the distance passed in the fourth second, notice that in the *first three* seconds it has passed nine spaces; that in the *next three* seconds it will go, by its velocity alone, eighteen spaces, and that in *one* of these three seconds, which would be the fourth second, it would go six spaces. Mark six spaces for velocity, and add one for the action of gravitation.

Fig. 38.

Construction of the Table. — Now, in this diagram the values of time, space, and velocity, stand clearly before us, and we may put them into a tabular form. In the first column, headed T, put the number of seconds, 1, 2, 3, 4. In the second column, headed S, put the total space passed over at the end of these seconds, representing the distance A B by d. In the third column, headed V, put the velocities gained at the end of each of the seconds. And, finally, in the fourth column, headed s, put the spaces passed in each separate second.

T.	S.	V.	s.
1	1d	2d	1d
2	4d	4d	3d
3	9d	6d	5d
4	16d		7d

From the Table obtain the Laws. — The relation between time, space, and velocity, may be seen by comparing their values given in this table.

Notice, FIRST, that the values of S have the same ratio as the squares of the value of T. For instance, take the spaces 4d and 9d, with the corresponding times 2 and 3, we find that $4d : 9d :: 2^2 : 3^2$. Hence *the spaces passed by a falling body in different times are as the squares of the times.*

Notice, SECOND, that the values of V have the same ratio as the values of T. Thus we find that the velocities 4d and 6d have the same ratio as 2 and 3, the corresponding values of time. Hence *the velocities of a falling body at the end of successive intervals of time will vary as the time of fall.*

Notice, THIRD, that *the spaces passed in separate seconds* (the values of s) *are as the odd numbers,* 1, 3, 5, 7, &c.

From the Table also obtain Formulas. — It will be seen, that by squaring any one of the values of T in the table, and then multiplying by d, the corresponding value of S will be obtained. Hence,

$$S = T^2 d. \quad (1.)$$

Again, we may discover that if the value of T in any case be multiplied by 2d, the corresponding value of V will be produced. Hence,

$$V = 2\,d\,T. \quad (2.)$$

We see, again, that if the value of S in any case be multiplied by d, the square root of this product, multiplied by 2, gives the value of V. Hence,

$$V = 2\sqrt{Sd}. \quad (3.)$$

Finally, a little attention will show, that if the value of T in any case be multiplied by 2, the product diminished by 1, and the remainder multiplied by d, the corresponding value of s will be obtained. Hence,

$$s = (2\,T - 1)\,d. \quad (4.)$$

Acceleration. — The velocity added in each interval of time is called the ACCELERATION. The acceleration is always represented by the letter g. The diagram shows that $d = \frac{1}{2}g$.

Let us put $\frac{1}{2}g$, which is the value of d, in place of d, and the four formulas become,

$$S = \tfrac{1}{2} g\, T^2. \quad (1.)$$
$$V = g\, T. \quad (2.)$$
$$V = \sqrt{2\,S\,g}. \quad (3.)$$
$$s = \tfrac{1}{2} g\, (2\,T - 1). \quad (4.)$$

The Value of g. — In all cases g represents twice the distance passed by the body in the *first* interval of time. Its value will be different for different forces. When gravitation is the constant force which causes the motion, the value of g is $32\tfrac{1}{6}$ feet.

By these Formulas solve Problems. — By the use of these four formulas, all problems in uniformly accelerated motion may be solved. A single illustration will show how they may be used. If a stone be dropped into a well whose mouth is $144\tfrac{3}{4}$ feet above the water, how long will it take to reach the water? Since gravitation produces this motion, the value of g is $32\tfrac{1}{6}$ feet. The $144\tfrac{3}{4}$ feet is the value of S, and the value of T is required. The relation between these elements is expressed by the formula $S = \tfrac{1}{2} g\, T^2$; and by substituting the given values we have $144\tfrac{3}{4} = T^2 \times 16\tfrac{1}{12}$. The value of T, from this equation, is 3 seconds.

SECTION II.

MOTION PRODUCED BY MORE THAN ONE FORCE.

28. If a body be acted upon by two forces which, separately, would cause it to describe the adjacent sides of a parallelogram, they will be equivalent to a single force, causing it to move through the diagonal of the parallelogram.

Hence the effect of two forces may be found by representing them by the two sides of a parallelogram, and then drawing the diagonal. (*A*. 111–117.)

If a Body be acted on by two Forces. — Forces seldom act singly. It is by the combined action of at least two, often of more, that almost every motion is produced. The action of two forces may be illustrated by a very simple experiment. Place a ball at one corner of the table. Snap it with the fingers, lengthwise of the table, and it will roll along the side; or snap it across the table, and it will roll across the end. But skillfully snap it both ways at the same time, using both hands for the purpose, and it will roll in neither of these directions, but will move obliquely across the table.

The same thing is true of the action of natural forces, such as wind and tide. A ship, driven north by a direct wind, may at the same time be drifted east by a tide moving eastward. If so, it will at every moment be moving north and east, or in a straight line toward the north-east.

Acting along the adjacent Sides of a Parallelogram. — The conditions of the motion of both the ball and the ship may be represented to the eye. Let A (Fig. 39) represent the original place of the ship. Suppose that while the wind, if acting alone, would move the ship to B, the force of the stream, if acting alone,

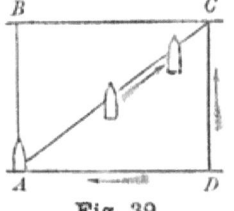

Fig. 39.

would move it to D *in the same time:* then, when both act at once, the body will neither go to B nor D, but will go along the diagonal line A C, and will reach the point C in the same time it would have taken to go to either B or D.

They are equivalent to a single Force. — The two forces, acting in the directions A B and A D, produce a single motion along the line A C. A single force acting in the direction of A C would have produced the same effect. Hence *two forces, acting in the directions of the sides of the parallelogram, are equivalent to a single force acting in the direction of the diagonal.*

The separate forces are called COMPONENTS; the single force which would produce the same effect is called the RESULTANT; and the process of finding the resultant is called the COMPOSITION OF FORCES.

The Resultant of two Forces may be found. — The resultant of two forces may be found by representing them by two adjacent sides of a parallelogram, and then drawing the diagonal. The lengths of the lines represent the strength, or the *intensity*, of the forces.

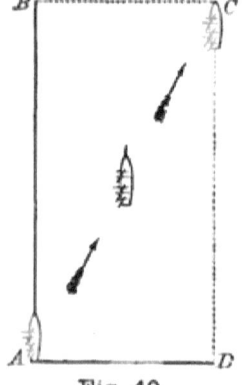

Fig. 40.

In the case of the ship, for instance, suppose the wind able to drive it ten miles, while the tide can drift it five miles. To find the actual path of the ship, draw the line A B (Fig. 40), to represent the ten miles, and then the line A D, at right angles to it, and *one-half as long*, to represent the five miles. Draw the lines B C and C D, to complete the parallelogram, and then draw the diagonal A C. This line represents the path of the ship, or the resultant of the two forces.

The Resultant of more than two Forces. — If more than two forces act at once, the resultant of all may be found by repeating the process. Find the resultant of two of them first; then compare this resultant and a third force; this second resultant and a fourth force; and so continue until all the forces have been used: the last resultant will be the resultant of all the forces.

29. Any force may be resolved into two others, which, acting together, would produce the same effect. This is done when we wish to know what part of a given force can be made available in a direction different from that in which it acts.

A Force may be resolved. — To find the components of a given force, we may represent it by a line, and make this line the diagonal of a parallelogram; the adjacent sides of this parallelogram will represent the components. More than one parallelogram can be drawn on the same diagonal; so more than one set of components may be found for a single force.

The process of finding the components of a single force is called the RESOLUTION OF FORCES.

To find the Component which acts in a given Direction. — When a ball is thrown obliquely against the floor, it acts upon it with less force than when thrown perpendicularly against it. But a part of the force will still be exerted perpendicularly against the floor.

To illustrate this important point, let a ball, A (Fig. 41), be thrown against the floor, striking it at C. We may let the line A C represent the force with which the ball is thrown. Now construct the parallelogram, by drawing the lines A B and C D perpendicular to the floor, and then A D parallel to it. The lines A B and A D represent the components of the force A C.

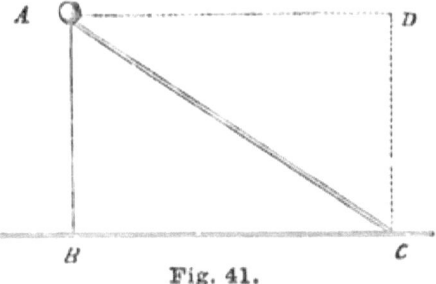

Fig. 41.

The line A B represents the amount of force exerted *per-*

pendicularly against the floor. To make the illustration more specific, we will suppose that, measuring the lines A C and A B, we find the latter to be $\frac{3}{5}$ as long as the former; if so, then the force exerted perpendicular to the floor will be $\frac{3}{5}$ of the force with which the ball is thrown.

To find the component which acts in *any* given direction, we may represent the original force by a straight line, and make it the diagonal of a parallelogram, one of whose adjacent sides is in the direction given. This side will represent the force required.

30. Two forces may act upon different points of a body in the same direction: their resultant will be equal to their sum.

The point of the body to which this resultant is applied will be as many times nearer to the greater force than the smaller one, as the greater exceeds the smaller in intensity.

The weight of a body is only the resultant of a set of parallel forces acting upon it in the same direction; and what is called the center of gravity is the point of application of this resultant. (*G.* 69; *A.* 202, 207, 210.)

Two Forces in the same Direction. — When two forces act upon a body in the same direction, they produce the same effect as a single force equal to their sum. If two horses, for example, draw a carriage, one with a force of two hundred pounds, and the other with a force of three hundred pounds, it is clear that a single horse exerting a force of five hundred pounds would produce the same effect.

The Point of Application. — But if a single force is to take the place of two others, and produce exactly the same motion as they would when acting together, at what point of the body shall it be applied?

Suppose the body represented by A B (Fig. 42), to be

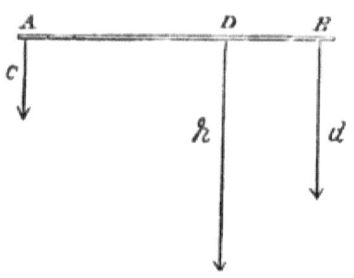

Fig. 42.

NATURAL PHILOSOPHY. 81

acted upon by two forces, represented by the lines c and d, one just half the length of the other, the lesser force being twenty-five pounds, the greater fifty pounds. Then the line r, just as long as both together, will represent the resultant, a force of seventy-five pounds. Now, if this resultant is to move A B exactly as the two components would, it must be applied at some point, D, as many times farther from A than from B as the force at A is times less than that at B. Since c is just half of d, the distance A D must be just twice as great as B D.

The Weight of a Body. — A body falling freely is an example of motion caused by the action of parallel compo-

Fig. 43.

nents. For, since the force of gravitation acts upon *every molecule* of the body, we may regard the entire force as made up of as many separate forces as there are molecules. The sum of all these components is their resultant, and the value of this resultant is the *weight* of the body.

The Center of Gravity. — The point of application of this resultant is the center of gravity. The center of gravity is usually defined to be *that point in a body, which being supported, the body will rest in any position*. One can balance a book on the tip of his finger: the tip of the finger must be exactly under the center of gravity of the book. This point being supported, the whole body will rest.

The center of gravity is the exact middle point of a body

of uniform density; it is toward the heavier side of one that is not.

In Fig. 43 the center of gravity in each body is at G.

What is the Line of Direction? — Now imagine a vertical line drawn through the center of gravity, as shown by the vertical dotted lines in Fig. 43. This line will show the direction in which the body would fall if it were left without support; and it is called the *line of direction*.

Principle of Stability. — That a body may stand upon a plane surface without falling, the line of direction must pass through its base.

Fig. 44.

One body stands more firmly than another, only because it is more difficult to throw its line of direction beyond its base. A load of hay is easily overturned, because, the center of gravity being high, the line of direction may be easily thrown outside the base. A load of stone, having no greater weight, stands firm, because, the center of gravity being low, the line of direction can with difficulty be thrown beyond its base.

Carriages may lean considerably to one side without overturning (Fig. 44); but an accident is sure to happen if they lean so far as to throw the line of direction beyond the lower side of the wheel.

NATURAL PHILOSOPHY. 83

Animals instinctively incline their bodies always in such a way as to keep their center of gravity over the space between their feet.

The showman offers a gold coin to the boy who will stand with his heels pressed against the wall of a room, and then pick it from the floor in front of him without falling. He is perfectly safe in making the offer. For no one can stoop without falling, unless, when he throws his head forward, he can, at the same time, throw some other part of his body backward far enough to keep his center of gravity over his feet. He can not do this with his heels pressed against a wall, and therefore can not win the coin.

31. Curved motion is produced by the action of at least two forces, one of which is a constant force, the other may not be.

The motion of a projectile is caused by the constant force of gravitation, and the impulse by which it is thrown. (*G.* 53; *A.* 159, 164.)

Curved Motion. — Whoever watches the varied and beautiful motions in nature, will find that they all take place in curves. In the ripples of the lake and the billows of the sea, he will see a wonderful variety of curved motions. The winds, and the clouds they carry, move in curves. Every swaying branch and leaf, and every nodding stalk of grass, moves in a curve.

Is produced by at least Two Forces. — The motion of a ball when fastened to the end of a string, and whirled around the hand, is an example of curved motion. It is produced by the action of two forces. The impulse of the hand H (Fig. 45), which starts the ball, would, if it could act alone, carry it in a straight line from A toward E. But the string H A, held firmly by the hand, is a constant force

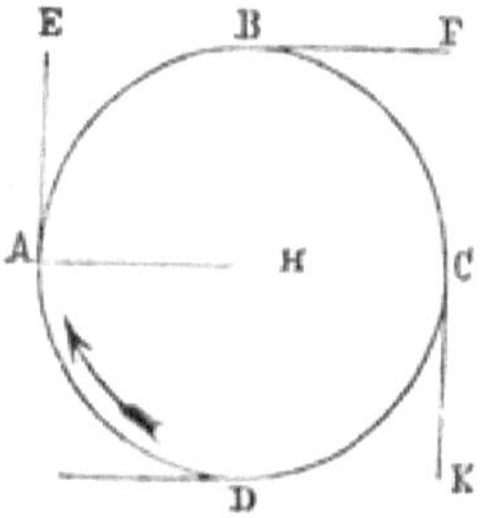

Fig. 45.

which pulls it away from that path. The resultant of these two forces is represented by the circumference A B C D.

One of which is Constant. — In the example just given, the force of the hand is an *impulsive* force; that of the string, a *constant* force; and a curved motion is the result. Two impulsive forces will cause motion in a straight line; two *equal* constant forces will do the same. Two constant forces that are unequal will cause a curved motion; one, at least, of the forces must be constant.

The Central Forces. — These two forces by which curved motion is produced are called the CENTRAL FORCES. One of them always acts toward the center of the circle around which the body is moving, and the other always in the direction of a tangent. The force which acts toward the center of the circle around which the body moves is called the CENTRIPETAL FORCE. The force of the string which holds the ball in its circuit is the centripetal force.

Let the centripetal force be destroyed, and the other central force will cause the ball to fly in a straight line, as B F or C K.

The most wonderful examples of the action of central forces are seen in the majestic movements of the heavenly bodies. Their orbits are ellipses. The impulse which drives the planets forward, and the attraction of the sun, are the central forces which hold them in their orbits.

Centrifugal Force. — The ball, when moving in the circle, is trying all the time to move in a straight line instead. This is the effect of its inertia. The string must overcome this inertia at every point, and this resistance of the ball is *a pull lengthwise of the string outward*. This resistance to deflection from a straight line is called the CENTRIFUGAL FORCE. This force takes no part in the production of the curved motion. It is simply the re-action of the moving body against the centripetal force.

Experiment. — A simple and pleasant experiment may be performed to illustrate the effect of centrifugal force. To the handle of a small pail, filled with water, tie a cord firmly. Grasp the cord, and swing the pail, fearlessly, in a vertical circle over the head; the centrifugal force will overcome the force of gravity, so that not a drop of water will fall, even when the pail is bottom side up over the head.

Illustrations. — Circus-riders incline their bodies toward the center of the ring around which they ride, that the centrifugal force may not throw them from their horses. Carriages, in very rapid motion around the corner of a street, are in danger of being overturned by this force.

Projectiles. — Any body thrown into the air is a projectile. The stone from the hand, the ball from the gun, and the arrow from the bow, are familiar examples of projectiles.

Their Motion is due to two Forces. — Leaving resistance of the air out of account, the motion of a projectile is due to the action of,

1st, The *impulse* which starts it on its journey; and,

2d, the *constant force* of gravity.

Range. — The horizontal distance is called the RANGE or the RANDOM of the projectile.

The range of a ball thrown from my hand is measured from my feet along the level ground to the spot where the ball strikes.

This distance depends upon the force applied to the projectile, and the angle at which it is thrown. Theory requires that the random be greatest when the projectile is thrown at an angle of 45°; but the resistance of the air very much modifies the motion, so that, in practice, the greatest range is obtained at an angle much below 45°. The greatest range of an arrow is when the angle is about 36°.

The science of gunnery rests upon the laws of projectiles. The most skillful gunner is he who can most accurately, under all circumstances, compare and combine the forces of gunpowder, gravitation, and the resistance of the air.

SECTION III.

ON THE MOTION OF LIQUIDS.

32. Water will issue from an opening in the side of a vessel with the same velocity which a body would gain by falling from the surface of the water to the center of the opening.

Hence the velocity of the jet of water will depend only on the distance of the orifice below the surface of the water in the vessel, and may be calculated by the formula, $V = \sqrt{2\,S\,g}$. (*A.* 346, 353–355.)

The Velocity of a Jet of Water the same as that of a Falling Body. — To prove this principle, we must remember: first, that water, confined in pipes, will rise as high as the source from which it comes; second, that a body thrown upward starts with the same velocity that it has when it gets back.

In Fig. 46, a bent tube, A, extends from near the bottom of a vessel of water. The water rises as high in the tube as in the vessel; it is the upward pressure of the water at A that pushes it up. The same *force* would be exerted on the water at A, if the tube were cut off at that point, and it would, if not resisted, throw the water to the *same height*, as shown on the other side of the figure, at B. But the velocity with which it must start from B, to reach the level of R, is the same it would gain by falling from that level back to B. If the tube were cut off at C, the water would issue under the *same*

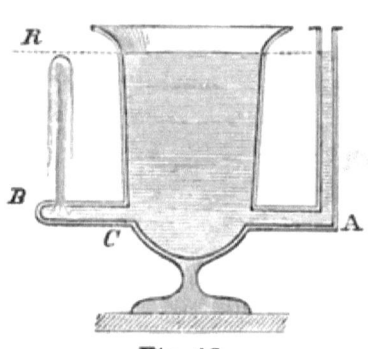

Fig. 46.

pressure, and, therefore, with the same velocity. Hence *the velocity with which the water issues is the same as that of a body falling from the surface of the water down to the center of the orifice.*

The Velocity of the Jet depends on the Distance of the Orifice below the Level of the Water. — The velocity of a falling body depends only on the height from which it has fallen. All bodies, whatever be their size or nature, fall with equal velocities. In the same manner, all liquids, however different in nature, will issue with equal velocities, if the openings from which they are thrown are at the same distance from the surface of the liquid in the reservoir.

Velocity calculated by the Formula, $V = \sqrt{2Sg}$. — Now, the velocity of a falling body is given by the equation $V = \sqrt{2Sg}$, and it is clear that the velocity of a jet of water will be given by the same formula, if S represents the distance of the orifice below the level of the water in the vessel.

If, for example, we would know the velocity of a jet of water from an orifice thirty-six feet below the surface in a reservoir, we put 36 for S in the formula. It then reads: $V = \sqrt{2 \times 36 \times 32}$. The value of V is 48; then the velocity of the water is 48 feet a second.

33. The quantity of water discharged from an orifice, depends upon its velocity, the size of the orifice, and the time of flow. It may be found by multiplying the values of these three things together.

To calculate the Quantity. — For example, how much water will flow from an orifice of $1\frac{1}{2}$ square feet area, at a depth of 9 feet below the surface of the water, in 10 seconds?

At a depth of 9 feet the water will issue with a velocity, $v = \sqrt{2 \times 9 \times 32} = 24$ ft. Now, if the opening were *one square foot*, then 24 cubic feet would issue in *one* second, and $24 \times 1\frac{1}{2} \times 10 = 360$ cubic feet must issue from the orifice of $1\frac{1}{2}$ square feet, in 10 seconds.

The rule is concisely expressed by the formula: —

$q = v \times a \times t$, in which

q represents the quantity of water discharged,
v represents the velocity,
a represents the area of the orifice,
t represents the time of flow.

In this equation there are four things, and it is clear that, any three of them being given, the fourth, whichever it may be, can be found. A single illustration will show how this is done.

Suppose 10,000 cubic feet of water must be discharged in 60 seconds, from an orifice so far below the surface of the water that the velocity of the jet is 250 feet a second: how large must the orifice be made?

In this problem, the value of v is given, 250 feet; the value of t is 60 seconds; the value of q is 10,000 cubic feet; the value of a is wanted. By putting the given values into the equation it becomes: —

$10,000 = 250 \times a \times 60$: hence,

$a = \frac{2}{3}$ of a square foot, or 96 square inches.

34. The velocity of a jet of water, and the quantity discharged, are found in practice to be much less than the foregoing theory would give. The actual amount may be increased by using short tubes of different shapes. (*G.* 210–213.)

The Quantity in Practice less than in Theory.— If we examine a jet of water flowing from an orifice in the side of a thin vessel, we shall see that it grows rapidly smaller, so that, at a little distance, its size is only about two-thirds as great as at the orifice. Beyond this point the contraction of the jet is gradual. The rapid contraction near the orifice is due to *cross currents*, caused by the water flowing toward the orifice from different directions in the vessel; these currents may be seen if there be any solid particles floating in the water. If the jet were the full size of the orifice, the quantity of water discharged would be what the theory gives; but, since it is only about two-thirds as large, there will be only about two-thirds as much water discharged.

The Quantity increased by using Tubes. — Short tubes inserted in the orifice are found to increase the actual flow. These tubes are either *cylindrical or conical*.

It is found that a cylindrical tube, whose length is not more than four times its diameter, if placed in the orifice, will increase the amount discharged to about .82 of that which theory gives. In this case the water *adheres* to the sides of the tube, and the tube is kept full, so that the contraction of the jet is prevented. By the use of conical tubes the amount discharged may be made still greater.

35. A liquid running freely down a vertical pipe exerts no lateral pressure.

A stream carries the adjacent air along with it.

These principles may be applied to produce either a blast of air or a vacuum. (*G.* 193, 194; *A.* 356.)

No lateral Pressure. — As the water falls, its velocity increases, and the stream must be smaller. It will not fill the pipe. The force of gravity is wholly expended in motion downward, and hence there is no pressure in any other direction.

Motion of the adjacent Air. — The power of a stream of water to drag air along with it is often shown when a faucet is opened. If the stream flows into a vessel partly filled with water, it penetrates to some depth, and sometimes, if the head is strong, it makes the water foam as if in violent ebullition. In any case an abundance of air-bubbles may be seen rising and breaking at the surface. This is the air which is dragged down by the stream. The adhesion of the two enables the stream to pull the air along.

Application to produce a Blast. — Let a small tube (V, Fig. 47) be inserted near the top of a vertical pipe. Let this pipe enter a bottle provided with two outlets, one near the bottom for the water to escape, the other near the top.

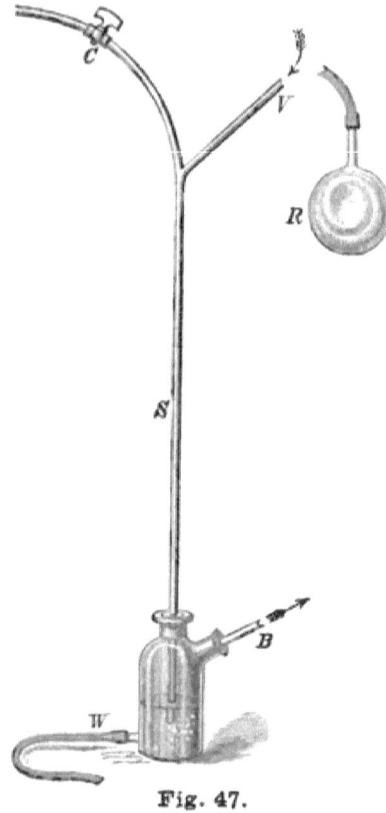

Fig. 47.

The stream of water in C S will drag the air down the vertical pipe, causing a current to enter through the small tube V, and to issue from the bottle in a steady blast from the jet B.

Application to produce a Vacuum. — Let the receiver R be attached to the small tube V, and the air in it will be taken out by the stream of water.

Various forms of apparatus for exhausting air, on this principle, are constructed. In the Sprengel pump, mercury is used. The tube is rather more than thirty inches in length, and the vacuum obtained is almost perfect. In Bunsen's pump, water is the liquid, and the tube is about thirty-four feet long to produce the best results.

SECTION IV.

ON THE MOTION OF AIR.

36. Air in motion is called wind. Winds are produced by the action of heat and the attraction of gravitation, upon the atmosphere; and, in case of the trade-winds, partly by the rotation of the earth on its axis. (*G*. 927–929.)

Wind. — The motion of air, called wind, is due to a difference in the temperature of two portions of the atmosphere. Heat expands air. One hundred cubic inches of hot air will *weigh less* than a hundred cubic inches of cold air. Then, if a portion of hot and light air is surrounded by that which is colder and heavier, it will rise, for the same reason that a cork rises in water. It will be pushed up out of the way by the heavier air, which takes its place.

Let us suppose that, in some particular part of the country, the air becomes heated more than in surrounding portions. This heated and lighter air will be pushed up by air *moving in* from all directions to take its place. This moving air is wind. People residing north of the heated place will observe a north wind, and those south of it a south wind.

Now, there is an unequal distribution of heat over the surface of the earth. It is caused partly by the changes of the seasons, and partly by various local causes. To it the *production* of winds is due. Their direction will be modified by many causes: the form of the surface over which they pass is an important one. As the same wind often blows in different directions on different sides of a house, or as blocks of buildings compel the wind to sweep up and down the various streets of a city, so the hills and valleys of a country, or the presence of forests or plains, will modify the direction of the winds that blow over them.

The Trade-Winds. — The trade-winds require particular notice. They occur in the equatorial parts of the earth, and *always blow in the same directions*. Over a surface of about 30° of latitude on the north side of the equator, they blow from the north-east toward the south-west; while south of the equator, over about the same width of zone, they blow from the south-east toward the north-west. These directions are maintained so constantly, that mariners count upon the trade-winds with almost the same certainty as upon the rising and setting of the sun.

Due to Heat and the Rotation of the Earth. — To

explain this phenomenon we must remember: first, that the equatorial region is constantly heated by the sun more than parts of the earth either north or south; and, second, that the earth revolves from west to east, the equatorial parts moving most swiftly.

The heated air at the equator, lighter than the air either north or south of it, will be pushed up, while currents of colder air from the north and from the south will move toward the equator. But the equatorial parts of the earth move toward the east more swiftly than other parts: the air from the north must, therefore, pass over portions of the earth which move eastward faster than itself, and it will be left behind. We find, then, that there is a real motion from the north, and at the same time an apparent motion from the east; these two motions combined make the direction of the wind to be from the north-east. A similar explanation will show why the southern trade-wind blows from the south-east toward the north-west.

SECTION V.

ON VIBRATION.

37. Examples of vibration are abundant among the motions of Solids, Liquids, and Gases.

Vibration is an alternate movement back and forth. (*G.* 56, 80.)

Examples. — If, with the finger, we sink one scale-pan of a balance, it will continue to pass alternately up and down over the same path for a long time after the finger is removed: it *vibrates*. Or if, instead of pushing it down, we pull the scale-pan to one side, and then release it, it will swing back and forth for a long time: this alternate motion, to and fro, is *vibration*. Suppose a ball, hung by a fine wire, be twirled by the fingers so as to twist the wire: let go of it, and, speedily untwisting the wire, it will go on for a time twisting it up the other way. The ball rotates, first in one direction and then in the other; and this alternate motion is *vibration*.

Or, take a bent glass tube; pour water into it until the arms are two-thirds full; tip it to one side, and then suddenly bring it back to a vertical positition. The water will rise and fall in the arms of the tube, and will continue this alternate motion up and down for some time. In this case a *liquid* vibrates.

Gases may be made to vibrate in the same way.

Definition.—All these motions are alike in one respect, different as they seem to be in every other. It is in every case a motion to and fro, or alternately back and forth. Motion of this kind is called VIBRATION.

I.—THE PENDULUM.

38. The pendulum vibrates under the influence of Gravitation and Inertia. Its vibration is governed by three laws:—

1st, The time of one vibration varies as the square root of the length of the pendulum.

2d, The time of one vibration varies inversely as the square root of the force of gravity.

3d, The time of one vibration is independent of the length of the arc through which the pendulum vibrates.

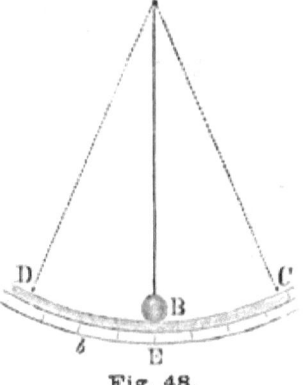

Fig. 48.

The Pendulum.—A body hanging from a fixed point under which it can swing freely is called a PENDULUM. In Fig. 48, the pendulum is represented as a ball B, hung from a point A.

If this ball be lifted from the point B to C, and then released from the hand, it will swing back and forth through the arc D C, going a less and less distance, until finally it will stop at B. Its motion from one end of its arc D, to the other C, is ONE VIBRATION; and the *distance* B C, through which it vibrates on either side of its place of rest, is called the AMPLITUDE of vibration.

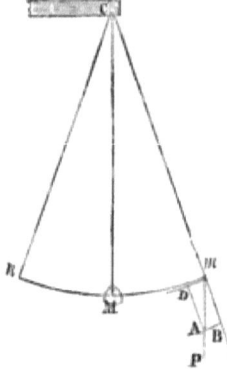

Fig. 49.

It vibrates under the Influence of Gravitation and Inertia. — Suppose a ball at M (Fig. 49) to represent a pendulum hung from the fixed point C, by a cord M C. Now, if this ball be lifted to the point *m*, and for a moment held there, the force of gravity will act upon it in a vertical direction. We will represent this force by the line *m* A, and resolve it into two components (see p. 79), shown by the lines *m* D and *m* B. The force *m* B acts lengthwise of the string *without effect to move the ball ;* the other force, *m* D, at right angles to the first, will pull the ball toward the point M. If the ball is allowed to fall to M, its inertia will carry it beyond that point; but gravitation will then be pulling it back with just the *same power* that it exerted to pull the ball from *m* to M. It will rise from M to *n*, a distance just as far from M, as it has fallen from *m*. It will there stop, and gravitation will bring it back to M, while its inertia will carry it up to *m ;* and if there were no resistance to its motion it would vibrate for ever through the arc *n m*. The resistance of the air, and the friction of the cord on the hook, will finally make it stop at M.

The first Law. — If two pendulums of different lengths (P and P', Fig. 50) be made to vibrate together, the short one will be seen to vibrate much faster than the other. We learn from this that the time of vibration *depends* on the *length* of the pendulum.

Now, let us make one pendulum P *just four times* as long as the other P'. With a watch in the hand, we can easily count the number of vibrations it makes in one minute; and sixty divided by this number shows how long it takes to make *one* vibration. In the same way we can find the time it takes the shorter pendulum to make *one* vibration. Doing this, we find that P takes *twice* as long as P' to vibrate once. The pendulum being *four* times as long, the time of vibration is *two* times as great. Hence *the time of one vibration varies as the square root of the length of the pendulum.*

Illustration. — The length of a pendulum to vibrate in one second is about 39.1 inches; to vibrate in two seconds, it must be *four times* as long; to vibrate in one-half a second, it must be *one-fourth* as long.

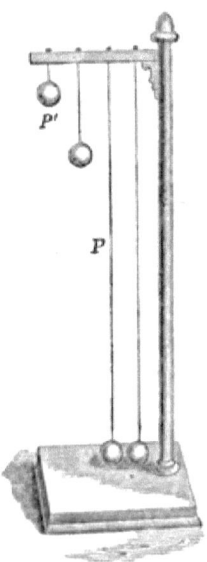

Fig. 50.

The second Law. — By calculating the force of gravity at different distances above the level of the sea, and then, by experiment, finding the time of one vibration made by the *same* pendulum at those places, it will be found that *the time of one vibration varies inversely as the square root of the force of gravity.*

The third Law. — Finally, if we make the pendulum P vibrate in a large arc, and find the time of one vibration, and then make it vibrate in a small arc, we shall find the time of one vibration to be the same. The pendulum must vibrate in equal times, no matter whether its arc be large or small. In other words, *the time of one vibration is independent of the arc through which the pendulum vibrates.*

This third law is absolutely true only when the arcs compared are *very small*. Yet, in the latitude of Paris, it is found that for a pendulum whose length is one meter, or 39.37 inches, the time of one vibration, through an arc of 8°, is only .000076 of a second longer than if its arc were infinitely small.

39. These laws apply to a single point in a pendulum, called the center of oscillation.

The Center of Oscillation. — The different molecules of a pendulum are at different distances from the point of suspension, and hence would vibrate in different times if they were not held together by cohesion. Although they are held together, and must all move at once, yet the forces that would make them vibrate differently are acting just the same as if they were not. The upper parts of the pendulum are *trying* to vibrate faster, and must be pulling the lower parts along; while the lower parts are *trying* to vibrate slower, and must be pulling the upper parts back. There must be some point in the pendulum at which these two struggles just balance each other. This point will vibrate just as fast as if it were influenced by no other molecules whatever. *A point in the pendulum which vibrates as if only under the influence of its own gravitation and inertia* is called the CENTER OF OSCILLATION. The center of oscillation is *generally* a little below the center of gravity of the pendulum-ball.

The Laws apply to this Point. — The three laws, obtained in the foregoing paragraph, apply to only this point, — the center of oscillation. Indeed, whenever we speak of the pendulum we refer to this point. By the length of a pendulum, we mean the distance from the point of support to the *center of oscillation ;* and, when we use the term *vibration*, we refer to the motion of this one point of the pendulum.

Huyghen's Discovery. — The centers of oscillation and suspension are interchangeable. If a pendulum be inverted, and suspended from its center of oscillation, its former center of suspension becomes its new center of oscillation, and its time of vibration is not changed.

40. There are several uses of the pendulum; we notice only two : —

1st, It is used to measure time.

2d, It is used to determine the form of the earth. (*G*. 82.)

Used to measure Time. — The vibrations of a pendulum are made in equal times. If, then, we know the time of one vibration, and can count the number made, we know the time during which the pendulum vibrates.

Now, the common clock is an instrument in which, by weights, friction and the resistance of air are overcome, so that the pendulum shall continue its motion, and by which the time of any number of vibrations is at the same time recorded by the hands moving over a graduated dial.

Used to determine the Form of the Earth. — The pendulum has been used to determine the shape of the earth. For this purpose, pendulums of the same length have been made to vibrate in different latitudes. It has been found that the time of one vibration is less and less as the pendulum approaches the poles of the earth. Now, to make the vibrations more rapid, the force of gravity must increase; and, if this force is stronger toward the poles, the surface of the earth must be nearer the center of the earth there than at the equator. The polar diameter must, therefore, be shorter than the equatorial diameter, and the shape of the earth must be that of an *oblate spheroid*.

II. — THE VIBRATIONS OF CORDS.

41. The vibrations of cords are due to the action of elasticity and inertia. They are governed by three laws: —

1st, The number of vibrations in a second varies inversely as the length of the cord.

2d, The number of vibrations in a second varies directly as the square root of the weight by which the cord is stretched, or its tension.

3d, The number of vibrations in a second varies inversely as the square root of the weight of a given length of the cord. (*G*. 259.)

The Vibration of Cords. — Let a cord or string be stretched between two fixed points (*a* and *b*, Fig. 51). By

Fig. 51.

taking hold of its middle point, the cord may be drawn to one side, *a e b*. Then loose it, and it will spring back, and go an equal distance on the other side, *a d b*, then return, and so continue to swing rapidly back and forth until it finally stops in its first position, *a c b*.

The motion of the cord from *e* to *d*, and back again, is a complete *vibration*. Its motion from *e* to *d* is a half-vibration, or, as generally called, a single vibration. The distance either side of its line of rest is the *amplitude* of vibration.

Due to Elasticity and Inertia. — When the force which stretches the string into the position *a e b* is withdrawn, elasticity moves it back to its first position, *a c b;* and the inertia gained by this motion throws it forward an equal distance, to *a d b*. The elasticity of the string again pulls it back to the position *a c b*, and its inertia carries it beyond; and thus, under the joint influence of elasticity and inertia, the string will swiftly vibrate, its amplitude growing less and less, on account of resistance, until at last it stops in its first position.

The Laws of Vibration. — The vibrations of cords are, in all cases, quite too rapid to be counted; and yet it will be impossible to establish any laws of vibration, unless we can find the *number* of vibrations made in a given time. How can this be done?[1]

However rapid the motion of the cord may be, the swiftness of electricity is yet greater: so by using electricity the cord may register the vibrations which it makes.

---

[1] The siren will be described in the chapter on sound: it seems desirable here to make the cord directly register its own vibrations, so that the *laws of vibration* shall stand independent of sound.

The apparatus devised for this purpose by the author is shown in Fig. 52.

The Electric Register. — The wire V is stretched over two bridges by a weight w. A strip of paper moistened with a mixture of potassium iodide and starch is drawn rapidly through the register R. Every vibration of the wire leaves a blue dot upon the paper, and the number of these dots made in a second can be easily counted.

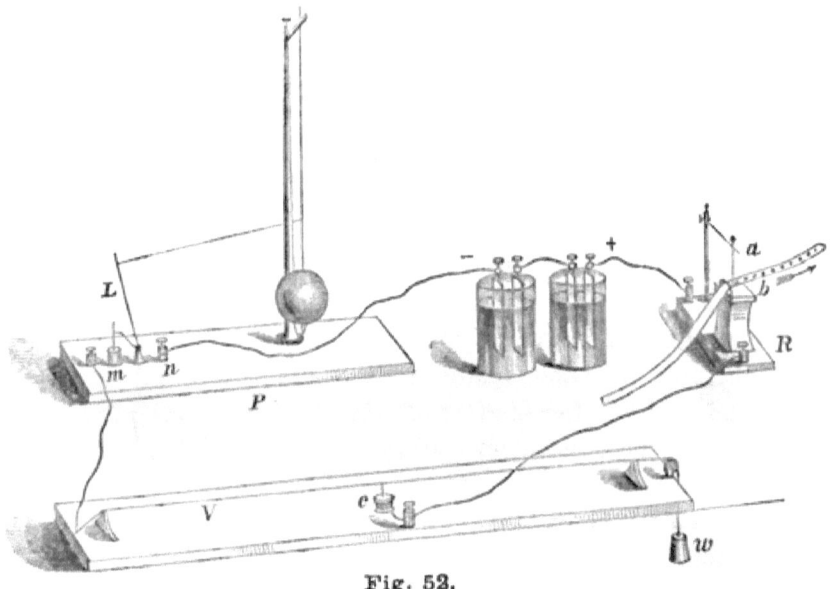

Fig. 52.

One needs to know something about electricity to clearly understand its action, and the complete description of the apparatus [1] should be left until after the study of that subject.

[1] The wire V, a time-measurer P, and the register R, are included in the same battery-circuit. The wire is enabled to open and close the circuit by means of a fine steel needle, fixed to its middle point, beneath which is a cup of mercury, c, the surface of which is so nearly in contact with the point at rest, that every vibration of the wire will immerse it in the liquid metal.

For the measurement of time a pendulum is employed, which holds the circuit during the time of one beat. The arrangement for this purpose is represented at P. A slender fiber is fastened to the pendulum-rod, and thence reaches over to the upper end of a light bent lever, L. This lever moves freely, and is in conducting communication with a binding post, n. Beneath the lower end of the lever is a mercury cup, m, in metallic connection with another binding post. When the

The first Law. — The wire, V, was taken 4 feet in length: stretched by a weight of 56 pounds at w, it made 315 complete or double vibrations in 3 seconds. The bridges were then placed under the wire so that the length of the vibrating part was 3 feet; it then made 420 vibrations in 3 seconds. But 315 is to 420 as 3 : 4. We see that when the lengths of the wire are as 4 : 3, the numbers of vibrations in the same time are as 3 : 4. Hence the number of vibrations in a given time varies inversely as the lengths of the wire.

The second Law. — The wire was again made 4 feet long, and the weight, w, 56 pounds. The vibrations in one second then numbered 105. When the weight, W, was changed to 14 pounds, the number of vibrations in one second was, in some experiments 52, and in others 53. The instrument can not register parts of a vibration; the true number is evidently between 52 and 53: we may call it $52\frac{1}{2}$. We see that when the weights are 56 and 14, or as 4 : 1, the numbers of vibrations made in a second are 105 and $52\frac{1}{2}$, or as 2 : 1. Hence the number of vibrations in a second varies directly as the square root of the weight by which the wire is stretched.

The third Law. — The wire which, being 4 feet long, and stretched with a weight of 56 pounds, gave 105 vibra-

heavy pendulum is at rest, the weight of the lever keeps the fiber tense; and the mercury surface is so adjusted as to be exactly in contact with the point of the lever, a most vital adjustment, but one very easily made. It will be seen that the pendulum, when vibrating, must compel the lower end of the lever to be alternately in and out of the mercury during the exact time of one vibration.

The record of the vibrating wire is made in the register R. A metallic point, a, presses upon a strip of chemically-prepared paper, which runs over a platinum surface, b. The pen, and the platinum beneath, are each provided with a binding post, by which it can be made a part of the circuit. The paper may be drawn through by the hand of the operator, the more swiftly, as the vibrations are more rapid.

Let the circuit be continuously closed while the paper is in motion, and a continuous colored line will be traced by the pen, a; but let the wire vibrate, and electric pulses in unison with it will traverse the paper, leaving a series of dots instead. If the pendulum be at the same time in motion, the pulses can traverse the paper only while the lever, L, is in mercury; and hence a *group* of dots on the paper will represent the vibrations of the wire in the unit of time.

tions a second, was found to weigh 19.4 grains to the foot in length. Another wire, weighing 43 grains to the foot, was taken of the same length and tension as the other; and the number of vibrations in one second was, in some experiments 70, and in others 71. The true number is between these: call it $70\frac{1}{2}$. Now, the weights of equal lengths of the wire being 19.4 : 43, the rates of vibrations are found to be 105 : $70\frac{1}{2}$; but 105 : $70\frac{1}{2}$ :: $\sqrt{43}$: $\sqrt{19.4}$, so nearly that we may infer that the number of vibrations a second varies inversely as the square root of the weights of equal lengths of the wire.

The Rate of Vibration is invariable. — Thus the rate at which a wire or cord may vibrate is fixed by the length, weight, and tension of the cord. Every piano-wire vibrates with a certain rapidity; and no human power can change it so long as the length, weight, and tension of the wire remain the same.

III. — VIBRATIONS OF OTHER BODIES.

Of a Bell. — In Fig. 53 we are shown a bell-shaped glass vessel with a little pendulum-ball hanging beside it. By drawing a violin-bow across the edge of this bell we make the glass vibrate; and we shall know that the vibrations are made, because the little pendulum-ball will fly back and forth with a violent clatter. The edge of the glass, springing back and forth, puts the ball in motion.

This is one of the many cases of vibration in which the motion is too delicate to be seen, and the existence of which would not be known if some way had not been discovered by which to make the vibrations show themselves. Cases of such invisible vibrations are very common. In fact, they already exist or may be produced in almost every solid body we can see around us.

Of Water. — Let the glass vessel (Fig. 53) be almost filled with water, and the bow then drawn across its edge. The fluid will be thrown into violent commotion. Hosts of little wavelets will be thrown up and down in quick succession upon its surface, the water being thrown into vibration by the vibration of the glass.

By skillfully drawing the bow these wavelets may be brought into four and sometimes into six beautiful groups separated from each other by portions of water which seem to be at rest. Not many effects as fine can be so easily produced.

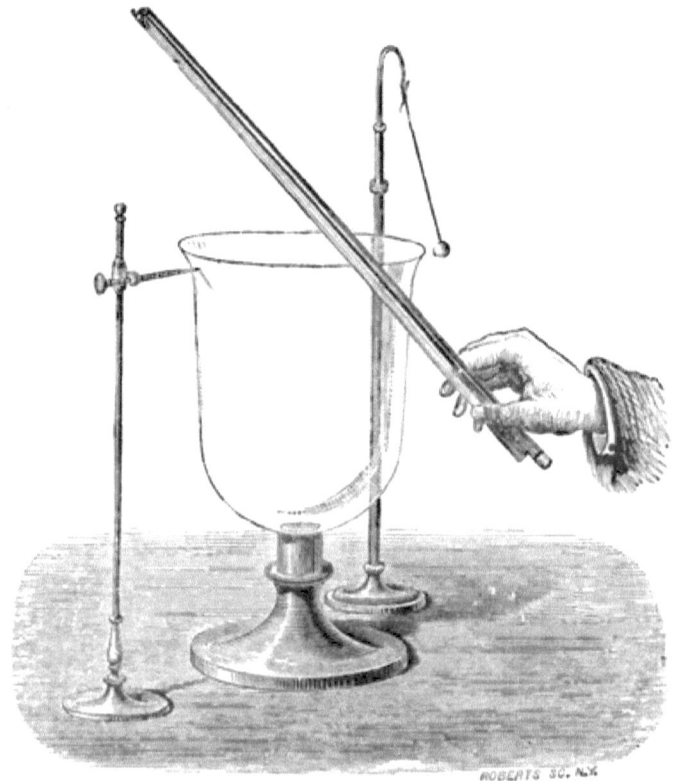

Fig. 53.

Of Air. — The air is so elastic that it yields to every force, even the very slightest, and then afterward springs back again. On this account it is in a state of vibration all the time. We can not stir a hand without causing the air to vibrate. It is made to tremble by every breath, and it quivers at every motion of our lips.

SECTION VI.

ON UNDULATIONS.

42. Vibrations may be transmitted from point to point in the same mass.

The motion then becomes an Undulation. (*A*. 475, 534.)

Transmission along a Cord. — Let a heavy cord, or better still, an India-rubber tube (A B, Fig. 54), several feet long, be fastened at one end to the wall or ceiling of the room. Take hold of the free end with one hand, and, by a sudden blow with the other, push the part B C aside, as shown in the figure. The little hillock thus formed will run swiftly up the tube to A, and then quickly down to the hand again. By carefully noticing the motion, it will be seen that while the hillock running up to A is on one side of the cord or tube, that which returns to the hand is on the other. Having gone to the top, as seen at A', it turns as seen at A'', and then comes down. Nor does it then stop: it will again and again run up and down the tube until, the height of the hillock growing less and less, it finally disappears.

The Motion appears to be Lengthwise of the Tube. — It is interesting and important to notice

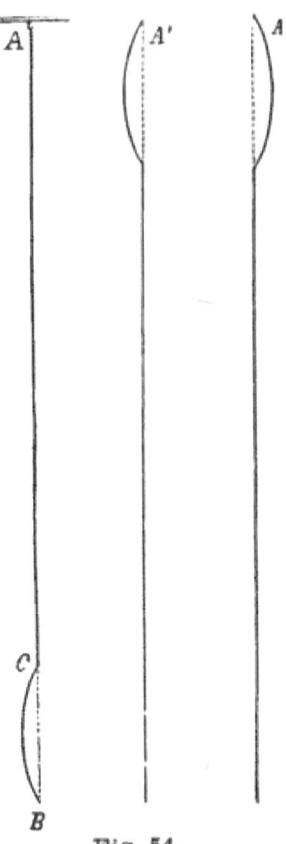

Fig. 54.

that while the motion appears to be lengthwise of the cord or tube, the only *real* motion of the parts is back and forth, across their first position. Each point of the cord is put in motion a little later than the one before it, and the *vibration progresses* along the cord.

Definition. — The *transmission of a vibration* through successive portions of a mass is an UNDULATION.

A Wave. — By starting several hillocks, one after the other quickly, the whole cord may be thrown into a series of hills and valleys, as shown in Fig. 55. In this case the motion between B and D, consisting of two parts, on opposite sides of the middle line, is called a WAVE. Two waves are represented in the figure.

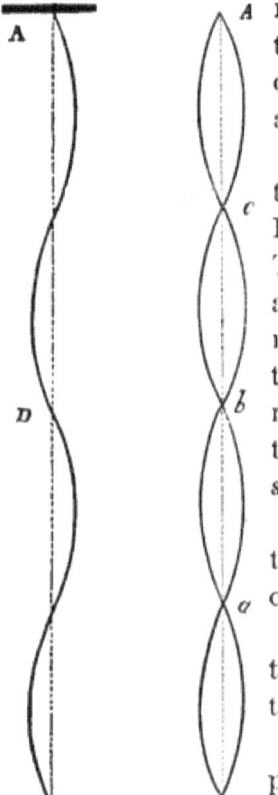

Fig. 55.   Fig. 56.

Phase. — When the cord is in motion as shown in Fig. 55, the two points B and D are moving exactly alike. They are moving in the same direction and with the same velocity. This is not true of any other points between these. Two points in a wave which move in the same direction and with the same velocity are said to be in the *same phase*.

Two points moving in opposite directions with the same velocity are in opposite phases.

Wave-Length. — The distance between two points in the same phase is the LENGTH of the wave.

Period. — The time required for any particle in the wave to make one *vibration* is the PERIOD. The undulation advances just one wave-length in the same time.

Interference of Waves. — By skillfully timing the impulses of the hand, the hillocks on both sides of the middle line in Fig. 55 may be made to turn themselves over at the same time. In that case, the tube will present the appearance shown in Fig. 56; the points *a*, *b*, *c*, being *almost* stationary, while the parts between are swinging to and fro across the middle line, making vibrations, just as if they were separate cords.

The points which appear to be at rest are called NODES, while the vibrating parts between them are called VENTRAL SEGMENTS.

At each node two waves meet, in opposite phases, one going up the cord, the other coming down. They pull that point *equally* in *opposite directions;* for this reason it remains at rest. Both pulses act in the segments also, but *not* with *equal strength;* and the cord moves in the direction of the stronger.

The mutual action of two or more undulations upon the same mass at once is called INTERFERENCE.

43. Undulatory motion is illustrated by Water-waves.

Two sets of water-waves may interfere with each other, and produce a single set different from either.

Water-Waves. — Let a pebble be tossed into the water of a lake or pond, and the tranquil surface will be carved into a series of circular ridges and furrows, which, growing gradually larger and larger, finally break against the shore. The motion appears to be in all directions outward from the pebble; but the little sticks and straws that may be resting upon the water at the time tell us, by their dancing, that the *real* motion of the water is, like their own, a motion only up and down.

A *wave* of water consists of two parts, — a ridge and a furrow.

Water-Waves may interfere. — Let two sets of water-waves be started at the same time, by dropping two pebbles at a little distance from each other. The two sets of growing circles very soon cross each other, and then the smooth surface of the water will be cut up into a curious confusion of dancing hummocks. Some of these hummocks will be twice as high as the ridges of either set of waves, while others will just lift their heads above the original surface of the water. When two sets of waves are thrown together, they are said to interfere.

But why are the hummocks of such different heights? It is clear that when two ridges come together their heights will be united, and the height of the hummock will be the *sum* of their separate heights. But when the ridges of one set enter the furrows of the other, the height of the resulting hummock will be equal to their *difference*. Now, as the waves are running across each other, the hummocks must be of various heights, limited on the one hand by the *sum* of the heights of the ridges of the two sets, and on the other by their *difference*.

The Principle of Interference. — *Any number of waves may traverse the same medium at the same time. They are equivalent to a single compound wave in which the motion at any point is the algebraic sum of the motions which the component waves impart.*

44. The undulations in air consist of alternate Rarefactions and Condensations. In free air the waves travel outward from their source in every possible Direction. Different sets must be constantly interfering. (*A*. 477, 478.)

Alternate Rarefactions and Condensations. — We have seen (see p. 44) how easily air may be compressed, and with what promptness it springs back to its former volume

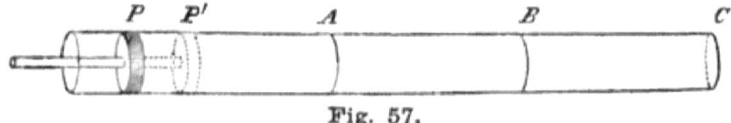

Fig. 57.

when the compressing force is removed. Now, suppose that near to one end of a long tube is a piston, P (Fig. 57). By suddenly pushing this piston forward to P', and then instantly pulling it back, the air in the whole length of the tube will be put in motion. Let us analyze this motion.

When the piston moves from P, it crowds the air before it; and, when it has reached P', this crowding effect will have gone forward to some point A, more or less distant. The space P' A is then filled with condensed air. When the pressure of the piston is removed, the condensed air springs back. It springs both ways, backward against the piston, and forward against the air at A. By its pressure against the air at A, the air in the space A B will be condensed. The next moment this air expands, and, pressing both ways, condenses the air B C in front of it, and also the air A P behind it. These two portions will, in this way, be condensed, while the air A B will be rarefied. The next instant these condensed portions spring back, and become rarefied, while the rarefied portion A B, and at the same time another part beyond C, will be condensed. The air is in this way thrown into a series of condensed and rarefied parts, alternately springing back and forth in the direction lengthwise of the tube. We need only add, that there is no *sudden* transition from condensed to rarefied air at the points A, B, and C. The mobility of air will not permit this. At the middle of the condensed part the condensation is greatest, while at the middle of the rarefied part is the greatest rarefaction, and between these points the change is gradual.

A *wave* of air consists of two parts, — a condensation and a rarefaction.

Waves in free Air go in all Directions. — The walls of a tube confine the air, and compel the waves to flow in the direction of its length. In free air the case is different. Every impulse by which the atmosphere at any point is suddenly condensed or rarefied is the center from which air-waves go outward in all directions.

Let a few grains of gunpowder be exploded. A little *sphere* of air at the point where the explosion occurs will be, for the moment, rarefied, while by its pressure a *shell* of air outside of it will be condensed. This condensed air, instantly springing back, condenses the air on both sides of it, and itself becomes rarefied. The waves will thus travel outward from the center, until the whole body of air in the room is thrown into a series of concentric *shells*, alternately condensed and rarefied.

How constant and complicated must be these vibrations of the air! Every sudden and local puff of wind, every forcible breath exhaled from the lungs; the fall of every stick and stone, — all these are the sources of as many different sets of waves spreading in all directions, darting across and through each other, too delicate to be seen or felt, but presenting to the mind a scene of activity far exceeding the power of the senses to appreciate.

Different Sets interfere. — Suppose two sets of air-waves come together: if their condensed parts coincide, a single set will be formed whose condensations are greater than either. If the condensed parts of one set coincide with the rarefied parts of the other, there will be a single set whose condensations are less than either. In the first case, if the two sets are equal, the resulting waves will be doubled; if, in the other case, the two sets are equal, they will destroy each other, leaving the air without waves.

Transverse and longitudinal Vibrations. — In the water-wave the water itself moves *vertically* while the wave moves horizontally. Whenever the parts of a medium vibrate at right angles to the direction of the wave, the vibration is Transverse.

In the air-wave the molecules of air move back and forth lengthwise of the wave itself. All such vibrations are Longitudinal.

SECTION VII.

REVIEW.

I. — SUMMARY OF PRINCIPLES.

Attraction and repulsion in one form or another determine the condition of rest or motion of masses and of molecules.

If a body could be acted on by a single force, its motion would be in a straight line unchanged for ever.

When forces act together, each does the same amount of work as it would if acting alone.

An impulsive force alone will produce uniform velocity.

A constant force alone will produce uniformly accelerated velocity.

Curved motion is produced by the action of two forces, one of which, at least, is constant.

The alternate action of two opposite forces produces vibration.

The transmission of a vibration from point to point in a body is an undulation.

II. — SUMMARY OF TOPICS.

23. Application of the fundamental ideas to explain the production of motion.

24. Newton's first law. — Second law. — Third law.

25. Velocity defined. — Uniform velocity. — An impulsive force. — Uniform motion due to an impulse. — Space=time × velocity.

26. A constant force. — Uniformly accelerated motion. — Difficulties in the way of experiment. — Overcome by Atwood's machine.

27. Experimental proof of the first principle. — Experimental proof of the second principle. — Analysis of the motion of a falling body by these principles. — Tabulate the results. — From the table obtain laws. — Also formulas.

— Acceleration defined. — The value of g. — Solution of problems by the formulas.

28. If a body be acted on by two forces. — Acting in the directions of adjacent sides of a parallelogram. — They are equivalent to a single force. — The resultant may be found.

29. Any force may be resolved. — To find the component in a given direction.

30. Two forces in the same direction. — Point of application. — The weight of a body. — The center of gravity. — The line of direction. — Principle of stability.

31. Curved motion. — Requires action of at least two forces. — One of which is constant. — The central forces. — Centripetal force. — Centrifugal force. — Experiments. — Illustrations. — Projectiles. — Range.

32. The velocity of a jet of water. — It depends on the distance of the orifice below the surface. — Calculated by a formula.

33. To calculate the quantity. — Example.

34. The quantity in practice less than in theory. — Increased by using tubes.

35. No lateral pressure by a vertical stream. — Motion of the adjacent air. — Applied to produce a blast. — Applied also to produce a vacuum.

36. Wind. — The trade-winds. — Due to heat and the rotation of the earth.

37. Examples of vibration. — Definition.

38. The pendulum. — Vibrates under the influence of gravity and inertia. — The first law. — The second law. — The third law.

39. The center of oscillation. — The laws apply to this point. — Huyghen's discovery.

40. The pendulum used to measure time. — To determine the form of the earth.

41. The vibration of cords. — Due to elasticity and inertia. — The laws of vibration. — First law. — Second law. — Third law. — Rate invariable. — Vibrations of a bell. — Of water. Of air.

42. Transmission of vibration along a cord. — The motion appears to be lengthwise. — Definition of undulation. — A wave. — Phase. — Wave-length. — Period. — Interference of waves.

43. Water-waves. — May interfere. — The principle of interference.

44. Alternate rarefaction and condensation of air. — Waves in free air go in all directions. — Different sets interfere. — Transverse and longitudinal vibrations.

III. — PROBLEMS.
Illustrating the Laws of Motion.

1. A body moves *uniformly* over a distance of 780 feet with a velocity of 5 feet a second : in what time did it go?
Ans. 156 seconds.

2. Under the influence of an *impulsive* force, a body moves at the rate of 25 feet a second: how far will it go in one minute?

3. A stone dropped from the top of a tower struck the ground in 4 seconds : how high is the tower?
Ans. $257\frac{1}{3}$ feet.

4. If the tower were $257\frac{1}{3}$ feet high, with what velocity would a stone strike the ground? *Ans.* $128\frac{2}{3}$ feet.

5. If the velocity of the stone should be $128\frac{2}{3}$ feet a second, how long a time had it been falling? *Ans.* 4 seconds.

6. A body falls 4 seconds : how far does it go in the fourth second? *Ans.* $112\frac{7}{12}$ feet.

7. Under the influence of a constant force, a body moves 3 *feet the first second:* how far will it go in 5 seconds?
Ans. 75 feet.

8. A body is falling toward the earth; it is at the same time moving horizontally under the influence of a constant force which made it go 10 feet in the first second: how far, horizontally, did it go in 8 seconds? *Ans.* 640 feet.

9. How far did it fall in the same time?
Ans. $1,029\frac{1}{3}$ feet.

10. With what velocity did it strike the ground?

Ans. $257\tfrac{1}{3}$ feet.

11. What velocity did it gain in a horizontal direction?

Ans. 160 feet.

12. How far did it go horizontally in the fifth second?

Ans. 90 feet.

13. How far did it fall in the fifth second?

Ans. $144\tfrac{3}{4}$ feet.

14. Under the influence of a constant force, a body goes 12 feet in the first 3 seconds: how far does it go in 18 seconds? Ans. 432 feet.

15. A ball is thrown directly upward, starting with a velocity of $96\tfrac{1}{2}$ feet: to what height will it rise?

Ans. $144\tfrac{3}{4}$ feet.

The motion of this ball thrown upward will be *retarded* by gravitation in exactly the same ratio that it is *accelerated* in falling to the ground again. The height to which it rises is the same as that from which it falls. This problem may be solved exactly as if the question were: From what height would the ball fall to gain a velocity of $96\tfrac{1}{2}$ feet a second?

16. A ball is shot upward with a velocity of 386 feet: how long will it continue to rise? Ans. 12 seconds.

17. How high does it go? Ans. 2,316 feet.

18. How long does it remain in the air?

Ans. 24 seconds.

19. How far does it rise in the *last* second of its ascent?

Ans. $16\tfrac{1}{12}$ feet.

20. How far does it fall in the *last* second of its descent?

Ans. $369\tfrac{11}{12}$ feet.

21. Suppose the large weights of Atwood's machine to be each $23\tfrac{1}{2}$ ounces, and the weight of the small bar to be one ounce. We find, by experiment, that the weight and bar go 4 inches in the first second: what is the acceleration by gravity? or, in other words, what is the value of g?

Ans. 32 feet.

In this case, the whole weight moved by the force of

gravitation on the bar is $23\frac{1}{2} + 23\frac{1}{2} + 1 = 48$ ounces. It is clear that 48 ounces will move only $\frac{1}{48}$ as far in one second as one ounce moved by the same force freely. Hence $4 \times 48 = 192$ inches would be the distance the bar would fall freely in the first second. This distance is equal to 16 feet. When the experiment is accurately made, at the level of the sea, it is found to be $16\frac{1}{12}$ feet. The acceleration is twice this distance.

22. An orifice is made near the bottom of a dam at a distance of 16 feet below the level of the water: with what velocity does the water issue? *Ans.* 32.

23. How large an orifice must be made in a dam 10 feet below the level of the water in order to supply 100 cubic feet of water a second?

24. Water issues from the muzzle of a hose directed upward with a velocity of 100 feet a second: to what height would it be thrown if there were no friction nor resistance of air?

CHAPTER IV.

ON ENERGY.

SECTION I.

ON DEFINITIONS AND MEASUREMENTS.

WHENEVER a body is put in motion, as when a stone is thrown from the hand, it will continue to move for a while, going more and more slowly, until it finally stops. To explain this fully, requires us to notice several things. Let us suppose it to be an arrow shot vertically upward (Fig. 58) : we must notice

The *Mass* of the arrow ;

The *Force* which throws it ;

The *Work* done in sending it to the top of its path ;

The *Energy* expended in doing this work.

Fig. 58.

45. Mass is the quantity of matter in a body. Its value is found by dividing the weight of the body by the acceleration caused by gravity.

That is, $m = \dfrac{w}{g}$.   (1.)

Mass and its Measure. — The quantity of matter contained in a body is always the same. Wherever the body may be upon the earth's surface, or even if it could be alone in space, it would contain the same quantity of matter; or, in other words, its *mass* would not change.

Not so with the *weight* of the body. Weight is caused by the earth's attraction, and must change as this attraction varies. The same body weighs less when the force of gravity is weaker, as in higher latitudes, for example; and if we imagine a body alone in space with none other to attract it, it has no weight at all. Clearly mass and weight are very different things.

But, if the weight of a body at any place is divided by the force of gravity at the same place, the quotient will be the same wherever the body may be, because the weight does vary exactly as the force of gravity varies. This quotient is, then, a numerical value which is the same for the same body anywhere, and hence it may represent the *mass* of it.

$$\text{Mass} = \frac{\text{weight}}{\text{gravity}} = \frac{w}{g}. \quad (1.)$$

46. Force is that which tends to produce, to destroy, or in any way to change, the motion of a body.

Its value is found by multiplying the mass of the body on which it acts by the velocity which it imparts.

That is, $f = m \times v.$ (2.) (*G.* 28, 29; *A.* 126, 127.)

Force and its Measure. — To move a mass of rock requires a certain amount of force. If the mass were twice or thrice as great, it would require twice or thrice as much force to move it with the same velocity, which shows us that *force varies directly as the mass* which it puts in motion.

But to move that same rock twice as fast would also need twice as much force; or, in other words, the *force varies directly as the velocity* which it produces.

The product of these two, mass and velocity, represents the force.

$$\text{Force} = \text{mass} \times \text{velocity} = m \times v. \quad (2.)$$

If a constant force is acting, then the *acceleration* is the velocity which is represented by v in this calculation.

The *force* which sends the arrow would be measured by the product of the mass of the arrow by the velocity with which it leaves the bow.

47. To overcome resistance, of whatever kind, is Work.

Its value is found by multiplying the weight of the body by the vertical height through which the force applied would lift it.

That is, $r = w\,s$. (3.)   (G. 61; A. 175, 177.)

Work. — The word "work" is used in science very much as it is elsewhere. It is applied to whatever requires an effort. In science its meaning is more precise than in common use. To carry weights up stairs is work, and if the weights are heavy we call it hard work. But in science we measure it, and give a numerical value to the work which is done.

The Unit. — Let us suppose each step of the stairs to be just one foot high. Now, if I lift a one-pound weight from the floor, and place it on the first step, I do a little work; and I should do exactly the same amount of work each time if I were to do it over and over again. Since I have lifted *one pound* to the height of *one foot*, we may call the little work done a FOOT-POUND. This is the unit in which work is measured. A foot-pound is the work done in lifting one pound to the height of one foot.

If I have carried the pound up ten steps of the stairs, each a foot high, how much work have I done?

Ans. 10 foot-pounds.

If I have lifted a weight of two pounds up one step, how much work have I done?   *Ans.* 2 foot-pounds.

If I have lifted two pounds up ten steps, how much work has been done?   *Ans.* 20 foot-pounds.

Rule. — To find the numerical value of work, multiply the weight lifted, by the height to which it is raised. Let work be represented by r, and height by s: then,

$$r = w \times s. \quad (3.)$$

The work done in throwing the arrow upward is found by

multiplying the weight of the arrow by the height to which it flies.

The French Unit. — In the French system the unit of work is called the *kilogram-meter:* it is the amount of work done in lifting a body weighing one kilogram to a height of one meter.

Work is independent of Time. — For, whether I lift my weights to the top of the stairs in one minute or in twenty minutes, the same amount of work is done.

To measure Work in other Directions. — But how can work be measured in foot-pounds *if the body is not lifted*

Fig. 59.

vertically? In Fig. 59 we see a cask of merchandise being drawn up the stairway. Perhaps it weighs two hundred pounds; and we ask how much work is done in bringing it to the top? It is easy to see that when it has been drawn from the floor to the platform along the steps, it has been lifted only through the vertical distance from the floor up to the platform on which the laborers stand. If the height of the platform above the floor is ten feet, then two hundred pounds lifted ten feet is equal to two thousand foot-pounds.

And this is the amount of work done in rolling the cask up the stairs.

And so, when work is done in any direction whatever, we may find to what vertical height the body would be lifted against gravity with the same amount of work; and then the weight of the body multiplied by this height is the amount of work done.

48. By energy we mean the ability to do work.

The energy of a body in motion is found by multiplying one-half its mass by the square of its velocity.

That is, $k = \dfrac{w\,v^2}{2\,g}$. (4.) (*G.* 63; *A.* 179–181.)

Energy. — When a nail is driven into wood, the moving hammer does the work. It is true, the ability to do this work is, at first, in the hand which wields the hammer; but the hand gives this ability to the hammer, and it is the hammer that finally drives the nail. The *hammer in motion* is able to overcome the resistance of the wood, and this *ability to do work* is its ENERGY. A heavier blow would drive the nail deeper, because the hammer would be endowed with greater energy.

A stone thrown vertically upward rises in opposition to the force of gravity. In throwing it to a greater height, more work is done; and, if you can project it higher than I can, it is because there is more energy in your hand than in mine.

Every moving body is endowed with energy.

Depends on Mass and Velocity. — A falling mass of rock possesses great energy; it shows it by crushing the obstacles in its pathway. A flying bullet also has great energy; it shows it by piercing whatever it may strike.

Now, the energy of the rock is not due to its great weight alone, for we know that it could overcome more resistance if it were moving *faster*. The energy of the bullet is not due alone to its great velocity; for, let a *heavier* ball fly with the same speed, and it would be still more destructive.

In these cases alike the energy depends on both mass and velocity. It depends on nothing else; and its relation to these two things is stated in the following

Law. — The energy of a moving body is proportional to its mass and to the square of its velocity.

A simple illustration may help us to see the truth of this law. Let a five-pound ball be thrown upward, starting with a velocity of 96 feet. That ball would rise to a height of 144 feet ($v = \sqrt{2\,g\,s}, \therefore s = \dfrac{v^2}{2g} = \dfrac{96^2}{64} = 144$), and the work done in lifting it so high would be $5 \times 144 = 720$ foot-pounds. And, if it were a ten-pound ball, the work would be $10 \times 144 = 1{,}440$ foot-pounds. Thus a *double mass* has the ability to do *twice as much work*.

Or let the five-pound ball be started with a velocity twice as great, 192 feet. Now, this velocity will carry the ball to a height of 576 feet, and the work done in lifting it so high is $5 \times 576 = 2{,}880$ foot-pounds; and this is *four times* as much as when the same ball started with a velocity of 96. Thus a *double velocity* gives the ability to do *four times the work*. In the same way *three* times the velocity would impart *nine* times as much energy to the moving body.

Relation of Energy to Work. — We have seen that the *time* in which any work is done is not taken into account in estimating work. To place a ton of merchandise upon a freight-car, the same amount of work is done, whether it is done in one hour or in two hours. But they who have done it in one hour are more fatigued than if they had spent two hours instead. To do work fast, requires more energy than to do it slowly. Time is thus seen to be an element in computing energy.

Comparison of Units. — The relation of energy to work may be seen again by comparing their units. The unit of work is the amount of work done in lifting *a one-pound weight one foot high*. The unit of energy is the energy expended in lifting *a one-pound weight one foot high in one second*.

To compute Energy. — In the formula $v = \sqrt{2gs}$, s is the height to which a body will rise if it start with a velocity represented by v. If you transform it a little, you will get $s = \dfrac{v^2}{2g}$. If, then, you know the velocity of a body, you can easily find how high that velocity would carry it if it were going vertically upward. You may square the velocity, and then divide by the value of $2g$, which is $64\frac{1}{3}$. For example: An arrow is shot horizontally with a velocity of 96 feet a second: how high would it have been able to rise, had it been shot directly upward? $96 \times 96 = 9{,}216$, and $9{,}216 \div 64\frac{1}{3} = 143+$ feet. The energy of the arrow would be able to carry it $143+$ feet high, and no more.

But if we can find the height to which a certain velocity will enable a body to rise, we can quickly find the work which it is able to do; for we need only multiply the weight by the height to find it in foot-pounds.

We see, then, that, in order to find how much work a moving body is able to do, we must *multiply its weight by the square of its velocity, and divide the product by* $64\frac{1}{3}$.

49. Energy is of two types, Kinetic and Potential.

Kinetic energy is the energy of a body due to its motion.

Potential energy is the energy of a body due to some advantageous position it holds. (*A.* 183–189.)

Kinetic Energy. — The arrow while at rest in the bow of the archer has no power to overcome resistance: it is without energy. In its flight from the bow, however, it has power to overcome the resistance of the air, and to pierce the target. The energy which it displays is *due to its motion*, and is called KINETIC ENERGY.

A body may have energy even when it is not doing work, as a man may possess money which he is not spending. If a body were falling in a perfect vacuum, it would meet no resistance, and hence could do no work. Still it would have the power to do work, and this is energy.

Potential Energy. — Again, a block of iron resting upon a high platform *has power* to overcome resistances. True, the platform prevents the exercise of that power, but the power is in it. Take away this hinderance, and the block falls, and when in motion expends the energy which was locked within it while at rest. The energy of the block upon the platform is *due to its position*, and is called POTENTIAL ENERGY.

The Advantageous Position. — The block resting upon the earth has no energy at all; lift it to the platform, and it is at once gifted with potential energy. The high position is the position of advantage. The block must be placed *at a distance* from the earth in order to be endowed with potential energy.

Another Definition. — Potential energy is energy *due to the separation of two bodies* which are acted upon by *a force tending to bring them together*.

50. There are several varieties of energy. When displayed in masses of matter it is called Mechanical Energy; in molecules it is called Molecular Energy; in atoms it is Atomic or Chemical Energy; in undulations, it is Radiant Energy; and in electrical actions it is Electrical Energy. All phenomena are only different exhibitions of these varieties of energy.

Mechanical Energy. — The energy of the train swiftly gliding along its iron track, of the waterfall and the tornado, are examples of energy displayed by masses, solid, liquid, and gaseous. A little thought will suggest many others.

Molecular Energy. — But, besides this energy of visible motion, there is energy of invisible molecular motion. We have reason to believe that *heat* is the energy of the molecules swiftly moving among themselves while the body itself is at rest.

It is well known, for example, that a nail, lying upon an anvil, may be heated until it can burn the fingers, by simply pounding it with a hammer. Now we suppose that the molecules of the nail tremble under the blows more and more as they are repeated, and that the heat we feel is the energy of this molecular motion spent against the hand.

Chemical Energy. — Not only masses and molecules, but atoms also, exhibit energy. Between the atoms of oxygen in the air and the atoms of carbon in coal, for example, there is a strong attraction. In the cold, this attraction can not bring the separate atoms together. They are in a condition like that of the earth and the block of iron which is kept from falling; their energy is *potential*. But if the atoms of oxygen and carbon can be allowed to *fall together*, as the block falls to earth, their energy becomes *kinetic*. This is what happens when coal burns in our furnaces or elsewhere.

Whatever power is obtained from fire is primarily due to the motion of atoms. It is the chemical energy in the fuel that drives the engine of the locomotive, the steamship, or the manufactory.

Radiant Energy. — When a cord vibrates, its energy is gradually taken away by the air until finally, when all is exhausted, the vibrations cease.

When a bell is struck, its molecules vibrate, but their energy is gradually given up to the air until exhausted, when the vibration ends.

By these and all other vibratory motions, the air is thrown into undulations; and this undulatory motion of the air possesses all the energy which was taken from the cord or the bell or any other vibrating body.

The energy exhibited by undulatory motion, in whatever medium it exists, is called *radiant* energy, because the undulations pass away from their source in all directions. They *radiate*. Radiant energy is that which is transmitted by undulations, outward in all directions from a center.

Electrical Energy. — The power of electricity to overcome resistance is shown in every thunder-shower. The lightning-flash is electricity in action, tearing a pathway through the air which resists its passage. And trees seen here and there, which have been torn and twisted by the lightning-stroke, display the terrible intensity of electrical energy.

SECTION II.

ON THE CONSERVATION OF ENERGY.

51. Energy, like matter, is indestructible. It may be transmitted from one body to another; it may be changed from one variety into another; but the sum total of energy can neither be increased nor diminished. This principle is known as the law of Conservation of Energy. (*G.* 65, 66, 484; *A.* 190–193.)

Energy may be transmitted. — In rolling marbles, when one strikes another it stops, and the other bounds away. The energy passes from one into the other. Place three or four in a row against the edge of a "ruler" to keep them in line, and then roll another violently against the end of the series. The ball at the other end bounds off, while the others remain at rest. The energy of the one you throw is *transmitted* from ball to ball until it reaches the last, which keeps it all, except what it gives up to the air and the table, which resist its motion.

Whenever a body in motion meets an obstacle, energy is transmitted to it.

Energy may be transmuted. — Let a ball be shot vertically upwards: on starting, its energy is kinetic. But it rises more and more slowly until it stops, and then falls again to the ground. Now, at the moment it stops at the top of its path, its energy is potential. During its flight upward, kinetic energy changes into potential energy; and again, during its fall, potential energy is changed into kinetic energy. There is no loss, but simply a change in the character of the energy.

Observe a swinging pendulum, and notice a similar transmutation of kinetic and potential energy.

Or take the case of the nail which is warmed by the blows of a hammer. The *mechanical energy* of the hammer is not lost when the hammer stops: it has been changed into *molecular energy*, or *heat*.

Whenever energy disappears, it has been transmuted into another form.

In the Exchange no Loss occurs. — The motion of a sledge-hammer can give rise to a certain amount of motion among the molecules of the anvil on which it falls, — no more, no less. The energy of the hammer is changed into molecular energy: it appears as heat. Now, if this heat could be all collected, and changed back into mechanical energy, it would be just sufficient to lift the hammer to the height from which it fell. The amount of heat produced by a given energy will always be the same, and when it disappears it can exert an energy just equal to that which caused it. Nature permits no loss in her exchanges.

The mechanical Equivalent of Heat. — Is it, then, possible to tell just how much heat will be produced by a given amount of mechanical energy, or how much energy a given amount of heat may exert? This has been done with the greatest accuracy.

The first Step in the investigation was, to settle upon some unit by which to measure the energy exerted, and another by which to measure the heat produced. The unit of energy chosen is the energy exerted by a weight of one pound, falling a distance of one foot. The unit of heat is the amount of heat required to raise the temperature of one pound of water $1°$ F.

Next, the question is, how many units of mechanical energy will produce one unit of heat? The honor of first answering this question is shared by Dr. Mayer of Germany, and Mr. Joule of England, who, at about the same time and by different methods, obtained results so much alike as to give impartial judges great confidence in, not less than admiration of, the labors of both. The experiments of Joule extended through seven laborious years. Dr. Tyndall describes them fully in his "Heat as a Mode of Motion." This author goes on to say, "It was found that the quantity of heat which would raise one pound of water one degree Fahrenheit in temperature is exactly equal to what would be generated if a pound weight, after having fallen through a height of 772 feet, should have its moving force destroyed by collision with the earth. Conversely, the amount of heat necessary to raise a pound of water one degree in temperature would, if all applied mechanically, be competent to raise a pound weight 772 feet high, or it would raise 772 *pounds one foot high.* The term 'foot-pound' has been introduced to express the lifting of one pound to the height of a foot." Then, 772 foot-pounds is what is called the MECHANICAL EQUIVALENT OF HEAT.

The three Essentials in the Law of Conservation.— The three characteristics of physical energy are:—

1st, It may be transmitted. — If one body gains, another loses.

2d, It may be transmuted. — If one form disappears, another takes its place.

3d, These changes are mathematical. — The quantity is unchanged.

The Law has not been successfully applied to social or mental Powers. — Work of the most exalted kind is wrought by the action of mental and social forces, but the law of conservation has not yet been extended over these energies. The mind which acts retains its mental energy without loss: *there seems to be no transmission.* It retains it the same in kind: *no transmutation has been discovered.* Moreover, in the present state of science *no unit of measure* for mental energy can be suggested.

SECTION III.

ON THE RECOGNITION OF ENERGY BY THE SENSES.

52. Molecular and radiant energy affects the organs of sense. Acting upon the sense of touch, it is recognized as Heat. Transmitted to the ear, it is recognized as Sound. By the eye it is recognized as Light. (*G.* 284, 417, 485.)

I. — HEAT.

Mechanical Action evolves Heat. — When the savage lights his fire by rubbing two pieces of hard wood together, he produces heat by *friction*. When, by repeated blows of a hammer, a nail is made too hot to handle, or when the iron-clad hoof of a horse " strikes fire " against a pavement-stone, heat is evolved by *percussion*. And, finally, when a piece of cold wood is heated by being squeezed between the plates of a hydrostatic press, heat is evolved by *pressure*. No two bodies can act upon each other, either by friction, by blows, or by sudden pressure, without evolving heat.

The Dynamic Theory. — It is now generally believed that heat is a kind of vibration, whose energy can affect the sense of touch. This theory supposes matter to be made up of molecules separated by definite distances; it goes further and supposes these molecules to be *in motion*, rapidly vibrating in the minute spaces between them. To increase this molecular motion, is to make a body hot; to lessen it, is to make the body cold. The theory assumes also the existence of the *ether*, which, according to the theory of light, must fill all space.

When we step from the shade into the sunlight, the gentle heat of its rays is instantly felt. The explanation is this: The molecules of the sun itself are in rapid vibration; they impart motion to the ether, whose undulations dart through the space between the sun and us, and, coming in contact with the person, impart their energy to the organ of the sense of touch, when we become immediately conscious of the presence of heat.

When bodies are heated by friction, their molecules are made to vibrate faster by the rubbing. Heat is evolved by percussion, because a blow increases the motion of the already trembling particles of the body struck. The same effect is produced by pressure.

II. — SOUND.

Hearing produced by Vibrations. — Let two books be clapped together, and every ear in the room receives a shock, to which the name of " sound " is given. The molecules of the books are made to vibrate by the blow, and these vibrations, acting upon the air in contact with them, produce air-waves. These air-waves, traveling outward in all directions, finally reach the ear; and the many parts of this organ, receiving the energy of these undulations, enable the mind to recognize the peculiar sensation which we call sound.

When we listen to the sound of a church-bell, we may in like manner imagine the molecules of the bell all in a state of tremulous motion, caused by the blows of the hammer. This motion causes undulation in the air in contact with the bell. The air-waves thus formed travel in all directions from the bell until they impart their energy to the ear.

The roar of a cataract is the result of vibrations caused by the falling water. The rolling sound of thunder is the effect of vibrations in air, caused by electricity. Every sound in nature, or that can be produced by art, may be traced back through the waves of some medium, to the vibrating molecules of a solid, liquid, or gaseous body.

Sound. — In the study of physics, the word " sound " refers to the undulations outside, rather than to the sensation in, the ear. Sound *is an undulatory motion in an elastic medium whose energy can affect the ear,*

III. — LIGHT.

Light is undulatory Motion. — It was once thought that light consists of minute particles of matter thrown in great abundance from the sun and some other bodies. It is now generally believed that light is a kind of undulation. But light will pass through the most perfect vacuum that can be made. Moreover, the atmosphere extends but a few miles above the earth, yet the light from the sun comes in floods through the vast distance which separates these bodies: what can there be between the sun and the earth, whose undulations bring to us the energy of sunlight?

The Ether. — Philosophers assume that there is a thin, elastic substance called ether, much finer and rarer than air, which fills all the spaces between the heavenly bodies, and enters into all those between the molecules of matter in every form. The undulations of this ether pass through a vacuum, through celestial spaces, through bodies like glass, and through the substances of the eye, until they strike the nerves of sight.

Vision. — When a gas-jet is suddenly lighted in a dark room, every eye present is dazzled by the brightness of the light. The explanation is this: The heated gas makes the ether in contact with it vibrate. This ether is between the particles of the air, and between the molecules of the substance of the eye. Its undulations, starting from the gas-jet, go through the air and into the eye; and, when these undulations of the ether impart their energy to the delicate nerves in the back part of this organ, we are made conscious of the presence of light.

Luminous Bodies. — Bodies which shine by their own light are called LUMINOUS bodies. They are bodies which can start undulations in the ether. Bodies which shine only by light which they receive from others are called NON-LUMINOUS bodies: they can not start undulations in the ether. The sun is a luminous body: so is a red-hot iron ball. All flames are luminous bodies. The moon is non-luminous. Almost all bodies on the earth are non-luminous: the light which they give to us is light which the sun first gave to them.

Light. — *Light is the undulation of the ethereal medium whose energy produces vision.*

Manifestations of Energy. — All the phenomena of heat and sound and light can be best explained by regarding them as only so many different exhibitions of molecular and radiant energy.

SECTION IV.

REVIEW.

I. — SUMMARY OF PRINCIPLES.

Force is that which in any way tends to affect the condition of a body as to rest or motion.

Work is the overcoming of resistance.

Energy is the ability to do work.

The unit of force is the force needed to move a unit of mass at the rate of one foot a second.

The unit of work is the work done in lifting one pound to a height of one foot.

The unit of energy is the energy expended in lifting one pound a foot high in one second.

To find the numerical value of a force, multiply the mass which the force moves, by the velocity with which it is moved.

This product, viz., $m \times v$, is called the MOMENTUM of the body. The momentum of a moving body is a measure of the force which is moving it.

To find the work done in lifting a body, multiply the weight of the body by the vertical height to which it is lifted.

To find the energy in a moving body, multiply the weight

of the body by the square of its velocity, and divide by 64$\frac{1}{3}$.

Kinetic energy is the ability of a moving body to overcome resistance.

Potential energy is the ability of a body at rest when in some high or advantageous position, to overcome resistance.

Energy of these two types is mechanical, molecular, chemical, radiant, or electrical, taking its name from the kind of work in which it appears.

All physical phenomena are only different exhibitions of energy.

PHYSICS IS THE SCIENCE OF MATTER AND ENERGY.

II. — SUMMARY OF TOPICS.

45. Mass and its measure.
46. Force and its measure.
47. Work. — The unit. — The French unit. — Rule for calculating work. — Work independent of time. — When the body is not lifted vertically.
48. Energy defined. — Energy depends on mass and velocity. — The law. — Illustrated. — Relation of energy to work. — Comparison of units. — To compute energy.
49. Kinetic energy. — Potential energy. — The advantageous position. — Another definition of potential energy.
50. Mechanical energy. — Molecular energy. — Chemical energy. — Radiant energy. — Electrical energy.
51. Energy may be transmitted. — Energy may be transmuted. — No loss of energy in the changes. — The mechanical equivalent of heat. — The three essentials in the law of conservation.
52. Mechanical action evolves heat. — The dynamic theory. — Hearing produced by vibrations. — Sound. — Light is undulatory motion. — The ether. — Vision. — Luminous bodies. — Light defined. — Manifestations of energy.

III. — PROBLEMS.

1. What FORCE is required to put a ball of iron weighing *one pound* in motion, with a velocity of *one foot* a second?
Ans. 1 unit.

2. What is the MOMENTUM of *one pound* moving with a velocity of *one foot* a second? *Ans.* 1 unit.

3. A locomotive weighing ten tons is to be put in motion with a velocity of thirty feet a second: what FORCE must the steam exert? *Ans.* 600000 units.

4. What is the MOMENTUM of a ball weighing ten pounds, and flying with a velocity of seventy-five feet a second?
Ans. 750 units.

5. How much WORK must be done to lift a bucket filled with water, weighing twenty pounds, from the bottom to the top of a well which is fifteen feet deep?
Ans. 300 foot-pounds.

6. A person whose weight is 120 pounds ascends a flight of stairs, reaching between two floors, which are 12 feet apart: how much WORK does he do?
Ans. 1440 foot-pounds.

7. Find the ENERGY of a body weighing five pounds, and moving at the rate of 45 feet a second.
Ans. 157.38+ foot-pounds.

8. What is the ENERGY of a locomotive whose weight is ten tons, and which is moving at the rate of thirty miles an hour? *Ans.* 601865.28+ foot-pounds.

9. Suppose the locomotive whose energy is found in the preceding problem should meet another having the same weight and velocity, what amount of energy would be expended to dash them to pieces?
Ans. 1203730.56 foot-pounds.

10. Suppose it were possible to convert the energy of the collision of these two locomotives into heat, how much water would the heat produced be able to raise from 32° F. to 212° F.? *Ans.* 8.66+ pounds.

CHAPTER V.

ON MOLECULAR ENERGY: HEAT.

SECTION I.

ON CONDUCTION AND CONVECTION.

53. Heat is conducted through some bodies much more freely than through others. Among solids the metals are the best conductors. Liquids are poor conductors, and gases still poorer. (*G.* 392, 394–398; *A.* 601, 602.)

Conduction. — Suppose that one end of a cold iron rod is held in the flame of a lamp. The heat will travel gradually from the flame through the rod, until the distant end becomes too warm to be held by the hand. The heat travels step by step from molecule to molecule to the end. This mode of transmitting heat is called Conduction.

Explanation. — If we would understand how the heat has made its little journey through the rod, we must picture to ourselves the delicate motion of the molecules of the iron. Those molecules in contact with the flame are made to vibrate; they swing against their neighbors, and put them also in more rapid motion; they, in turn, give motion to the next, and these to the next, until those at the distant end of the rod have finally received the shock. The energy of these molecules of the rod is intercepted by the hand in contact with them; the delicate nerves of touch receive the impulses, and announce the pain.

Definitions. — Some bodies conduct heat more freely than others. Those which conduct heat freely are called

Conductors; those which hinder its passage much are called Poor Conductors; and those which nearly or quite forbid its passage are called Non-conductors. The property in virtue of which bodies conduct heat is called Conductivity.

Metals are good Conductors. — Among solid bodies the metals, as a class, are the best conductors, but among metals there is great difference in conducting power. By a very simple experiment this may be illustrated. Plunge two spoons, one of silver and the other of German silver, into the same cup of hot tea: it will be found that the upper end of the silver spoon will become hot much quicker than that of the other. Among the best conductors we find silver, copper, gold, brass, tin, and iron, in the order named.

The Conductivity of Liquids. — The conducting power of liquids is very feeble. Water, for example, in a glass tube, with ice at the bottom, may be boiled without melting the ice, by applying the heat to the top of the water, or near the upper end of the tube. (Fig. 60.)

The Conductivity of Gases. — Whether gases conduct heat in the least degree, is doubted. Dry air is surely among the poorest conductors; and so, likewise, are all porous substances in which large quantities of air are inclosed.

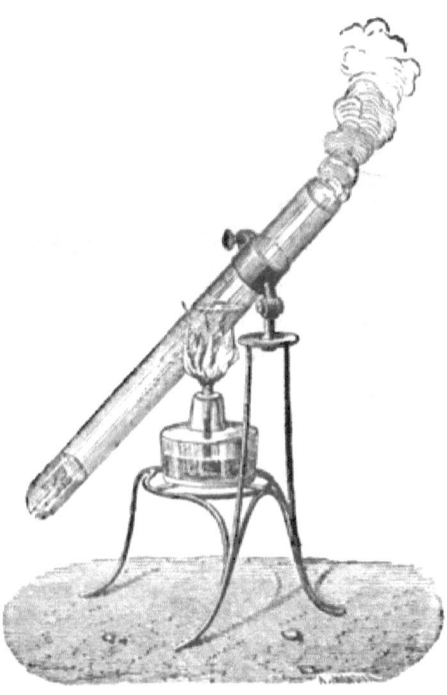

Fig. 60.

54. Convection takes place in bodies whose particles are free to move. Air is heated in no other way. Liquids are also heated by convection: solids never are.

Convection. — Heat is transmitted by convection when it is carried from place to place by moving particles of matter. The following very simple experiment will make this definition clear. Upon a plate of thick glass or a smooth block of wood put a bit of candle, lighted, and over it place a lamp-chimney so that its edge may project a little beyond the edge of the block (Fig. 61). If the edge of the chimney fits closely upon the top of the block, so that no air can enter

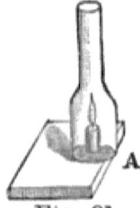

Fig. 61.

except at the open part A, the flame will flutter violently, showing that air is forced against it. If some light substance, such as down or cotton, be hung from a thread above the top of the chimney, it will be lifted away, showing that air is rising out of the chimney. Now, we know already that air is expanded by the heat; and we learn from this experiment that the cold air going under the glass pushes the expanded air away from the flame, up and out at the top of the chimney.

What we have seen in this experiment really takes place whenever a hot body is surrounded by colder air. The air in contact with a hot stove, for example, is heated and expanded. The colder air then pushes it away, and takes its place, only in turn to be heated and pushed away by other colder portions. The air goes to the stove, becomes heated, and moves away to other parts of the room, *carrying the heat with it*. This *transfer of heat by the moving particles* is called CONVECTION.

Air is heated in no other Way. — Air is heated only by convection. The heat of a stove does not go out to distant parts of a room to warm the air: *the air must go to the stove to be warmed.* So, too, the atmosphere is warmed by convection. The sunbeams coming through it do not warm it: they only warm the earth beneath it. Nor does the heat of the earth pass from particle to particle of the air, as it may in solid bodies: the heat of the earth warms only those particles in immediate contact with it. These rise and carry their heat with them to upper regions, while colder ones take their places in contact with the ground to get warm in turn, and then ascend.

Liquids heated by Convection. — Liquids are also heated by convection. A simple experiment will illustrate the convection of water. Into a flask or bottle of water put a little cochineal. Its particles are just about as heavy as those of the water, and will show by their motion whether there are currents in the water. Warm the bottom of the bottle, and the heated water will be seen to be rapidly leaving the bottom, while other portions are moving downward to take its place.

No Convection in Solids. — Solids can not be heated by convection, simply because their particles are not free to move among themselves.

SECTION II.

ON THE EFFECTS OF HEAT.

55. The action of heat is twofold: it raises the temperature of a body to which it is applied, and at the same time expands it. This is true of its action upon Solid, Liquid, and Gaseous bodies. (G. 303, 304, 309, 310; A. 660.)

The Action of Heat is Twofold. — That the temperature of bodies is raised by the application of heat, is too familiar to need illustration. That, while the temperature rises, the body grows larger, is known by such facts as the following: A ball of metal, which, when cold, just fits a ring, will be too large to enter it when hot. A clock-pendulum is longer in summer than in winter. The tire of a carriage-wheel is put on while hot; on cooling, it contracts, and binds the parts of the wheel firmly together.

The Expansion of Solids. — The expansion of a solid by heat may be shown by an experiment. A ball of iron or of brass is taken, just large enough to pass easily through a ring of the same material. The ball is then heated by a lamp, after which it will be too large to go through the ring. It will rest upon the ring (Fig. 62) until it is cold again, when it once more passes easily as at first. We see that heat makes this ball larger. And it has the same effect upon other solids.

Different solids expand unequally for the same increase of temperature; but each solid expands uniformly, — that is to say, two, three, or ten degrees of heat will produce respec-

Fig. 62.

tively two, three, or ten times as much expansion as one degree.

Expansion of Liquids. — To show the expansion of a liquid a glass bulb with a long open stem is used. The bulb is filled with water, and the stem partly filled, after which, if the bulb is plunged into hot water (Fig. 63), the water in the stem will be seen slowly rising.

Illustrations. — A kettle very nearly full of cold water will be quite full when the water is heated: the water will run over long before it boils.

Twenty gallons of alcohol in midwinter will become about twenty-one gallons in midsummer.

The Expansion of Gases. — The expansion of gases is more nearly uniform than that of either solids or liquids, and much greater for the same addition of heat. What is more remarkable is the fact that they *all expand at the same rate.* If we have 491 cubic inches of air at a temperature of 32° F., and add one degree of heat, there will be 492 cubic inches : it expands $\frac{1}{491}$ of its bulk. All gases expand at the same rate, $\frac{1}{491}$ of their bulk at 32°, for every additional degree. This fraction ($\frac{1}{491}$) of its volume at 32° F., which a gas expands for every degree of temperature, is called the Coefficient of Expansion for gases.

Fig. 63.

56. Temperature is measured by the expansion which accompanies it, in instruments called thermometers. There are three varieties of thermometers in use, — the Fahrenheit, the Centigrade, and the Réaumur. The air-thermometer is used to show delicate changes of temperature. (*G.* 287–293.)

Temperature measured by Expansion. — We have found that temperature and expansion increase at the same time by the addition of heat. Moreover, in the same body a certain amount of expansion *always* occurs with the same increase of temperature. By seeing the expansion we may therefore judge of the increase of temperature.

The expansion of solids is too slight, while that of gases is too great, to be conveniently used to measure the changing temperature of the air and other things. *Mercury* is a liquid metal, whose expansion is remarkably uniform, and neither too great nor too little for practical purposes. All common thermometers are made with it.

The Thermometer. — The mercurial thermometer consists of a glass tube terminating at one end in a bulb, and sealed at the other. The bulb and lower part of the tube are filled with mercury, the space in the tube above the mercury being a vacuum. Behind the tube is a graduated scale to show the height of the column of mercury.

There are three modes of graduating the scale, and this gives rise to three varieties of mercurial thermometer.

Fahrenheit. — In the *Fahrenheit* thermometer, the zero of the scale marks the height of the mercury in the tube when the bulb is placed in a mixture of snow and salt. When the bulb is put into boiling water, the mercury in the tube runs up to a point which is marked 212 on the scale. The distance between these points is divided into 212 equal parts called degrees, and this graduation is carried above and below these points. According to this thermometer, water boils at 212°, and freezes at 32°.

Centigrade. — In the *centigrade* thermometer, the zero point marks the height of the mercury in the tube when the bulb is placed in freezing water. The height to which it rises when the bulb is put into boiling water is marked 100, and the distance between these points is divided into 100 equal parts. The boiling point of water is, therefore, 100°; its freezing point is 0°.

Réaumur. — In the *Réaumur* thermometer, the zero marks the freezing point of water; the boiling point is marked 80.

Below Zero. — Degrees of temperature below the zero

point are generally indicated by the minus sign (−) placed before the number. Thus, −40° means a temperature 40° below zero.

Other Forms. — Mercury freezes at about −39° F. Temperatures below this point are measured by thermometers containing *alcohol*. Mercury boils at 660° F: temperatures above this point are measured by the expansion of *solid bodies*.

When it is necessary to show very delicate changes of temperature, the *air-thermometer* is used. This instrument has a variety of forms, but it consists essentially of a glass tube, terminating at one end in a bulb, the other end being open, and inserted into a cistern of colored liquid. (See Fig. 64.) The liquid fills a part of the tube: the rest of the tube and the bulb above is filled with air. A graduated scale is placed behind the tube. The air expands or contracts with every change of temperature, and accordingly drives the colored liquid down, or allows it to rise in the tube. The motion of the liquid shows the change in the temperature.

57. Temperature is the manifestation of *kinetic* energy; expansion, of *potential*. The first affects the touch, and is called SENSIBLE HEAT. The second does not affect the touch, and is called LATENT HEAT. (*G.* 434, 435, 438, 439, 443; *A.* 670.)

Fig. 64.

Rapidity of Motion, and Change of Position. — He who has a clear idea of the molecules can distinctly imagine the multitude of these little bodies of which any larger body is made up, separated from each other by minute distances and in rapid motion. Now, heat can make them vibrate faster; it may also push them farther apart, or otherwise change their position: it can do nothing more. Then, when heat is being applied to a bar of iron, let the mind picture to itself these two effects: the molecules of the bar vibrating more and more swiftly, and at the same time being pushed farther and farther apart. The first of these effects is manifested as temperature, the second as expansion.

In the first we have *molecules in motion*, or kinetic energy. In the second we have *molecules separated*, or potential energy.

Sensible and Latent Heat.—The heat which is expended in raising temperature is called Sensible Heat: it can affect the sense of touch. That which is used to produce expansion or change the positions of the molecules of a body is called Latent Heat: it does not affect the sense of touch. Now, the heat that goes into, or acts upon, any body, is divided into these two portions; one part sensible, the other latent.

Specific Heat.—But different substances do not divide it alike; that is, if the same amount be added to two substances, one of them will devote more of it to temperature, and less to expansion, than the other.

Let equal weights of water and mercury be placed over the same source of heat. The water divides the heat it receives into two parts, one to raise its temperature, the other to expand it. The mercury, receiving the same amount, divides it into two parts devoted to the same purposes, but the heat devoted to temperature is more than in water, while that devoted to expansion is less. We find that the temperature of mercury rises much faster than that of water: it takes thirty times as long to raise the water to a given temperature as it does the mercury. If it take thirty times as long, and one receives heat as fast as the other, there must be thirty times as much heat in the water as in the mercury when that temperature is reached by both. We see, then, that at the same temperature different substances may have very different quantities of heat in them. The relative quantities of heat in different bodies, at the same temperature, is called Specific Heats.

Water is the Standard of Specific Heat.—At a given

temperature water contains more heat than any other known substance. Its specific heat being 1, the specific heats of all other substances are fractional. The specific heat of mercury is .03. By this is meant, that, when equal weights of mercury and water are at the same temperature, the mercury will contain only .03 as much heat as the water.

58. The expansion of a solid body will continue nearly uniform until its temperature has reached the melting point. The temperature then stops rising, while the expansion increases and continues until the solid is melted. (*G.* 327–329, 335, 336.)

The Melting Point. — The temperature at which a solid body begins to melt is called its MELTING POINT. At this temperature, the repulsive force of heat nearly balances the cohesion of the molecules, and enables them to move freely among themselves. The body becomes a liquid, and the change from the solid to the liquid form is called LIQUEFACTION. The melting-point for different substances is not the same. Ice melts at 32° F.; mercury at —.39°; iron at about 3000°, and platinum at about 5000°.

The Temperature stops rising. — If heat be applied to a vessel of ice at 32°, the ice will melt, and the water formed will have the same temperature, 32°. So, too, when wax or iron or lead is melted, the liquid will have the same temperature as the solid which is melting.

But the Expansion increases. — But in the case of all the substances above named, except ice, the expansion is greater at the melting point than before it was reached. The liquid fills more space than the solid from which it was formed. It should be so, because the heat is *all* expended to change the position of the molecules, whereas, before, a part of it was used up to produce a rise of temperature.

An Exception. — Ice contracts when melting; the water formed fills less space than the ice. In this case, likewise, all the heat applied is expended in *changing* the *position* of the molecules, but not in pushing them farther apart, for they occupy less space than before the change occurred. The change consists in throwing the molecules out of their crystalline arrangement. The water will continue to contract until it reaches a temperature of 39°, after which it expands.

Those who have attempted to melt ice or snow, for domestic purposes, remember how slow the process is. The amount of heat required to simply melt the snow, without making it any warmer, is very great; the same amount applied to the water formed would raise its temperature 142°. This is the amount of heat which is used up in changing the relative positions of the molecules, and becomes latent. Hence the *latent heat of water* is said to be 142°.

59. The expansion of a liquid will continue gradual until the Boiling Point is reached, a temperature depending upon the Purity of the liquid, upon the Nature of the vessel in which it is heated, and upon the Pressure it sustains. At the boiling point, the temperature stops rising, while the expansion greatly increases, and continues until the liquid is vaporized. (*G.* 351, 352, 354, 358; *A.* 684.)

The Boiling Point. — The temperature at which a liquid begins to boil is called its BOILING POINT. At this temperature, the body rapidly becomes a vapor, and the change is called VAPORIZATION.

Evaporation. — Liquids do indeed change to vapor at all temperatures. Even from freezing water, more or less vapor is ever slowly rising. This slow change is called EVAPORATION. The boiling point for different liquids is not the same. Water boils at 212° F., alcohol at 173°, ether at 95°. The boiling point for the same liquid is not always the same: it depends upon three circumstances.

The Boiling Point depends on the Purity of the Liquid. — It is affected, first, by the presence of impurities in the liquid. The presence of some impurities raises the boiling point; of others, lowers it. Salt water, for example, boils at a higher temperature than pure water, while that which contains air boils at a much lower temperature than that which contains none.

It depends on the Nature of the Vessel. — In an iron vessel, water will boil at a lower temperature than in one of glass. It is so because there is a stronger adhesion between water and glass than between water and iron. The stronger adhesion requires a stronger heat to overcome it.

It depends on the Pressure. — But the most important circumstance on which the boiling point of a liquid depends is the pressure it sustains. This pressure is due to the atmosphere, to the weight of the liquid itself, and to any force which may be brought to bear upon it by artificial means. Whatever may be its cause, the effect of pressure is to raise the boiling point. It is well known that water boils at a lower temperature on the top of a mountain than at its base. It does so because the pressure of the air upon it is less.

This very important principle may be easily illustrated by experiment. For this purpose take a glass flask, or, better, a bolt head (Fig. 65), and put into it water enough to fill the stem and a small part of the bulb. Invert it so that the water may be boiled by holding the bulb over the flame of a lamp. Boil it until the steam issues freely from the stem, and then, removing it from the flame, cork the stem at the same instant. The air has been driven out from the instrument, and nothing remains, to press upon the water, but steam. Turn the bulb upward so that the water may fill the stem; pour *cold* water upon the bulb, and the water inside will boil violently. Even when the tube has become so cold that it may be handled without inconvenience, a fresh bath of cold water will cause the boiling to continue. It boils at the low temperature because the cold water, condensing the steam, removes the pressure from its surface.

Fig. 65.

The Temperature stops rising. — No matter how much heat may be applied to boiling water, its temperature is not raised. Moreover, the temperature of the steam is always that of the water from which it is made.

By the following experiment these facts may be illustrated. Water is placed in an open vessel V (Fig. 66). Into the water is plunged the bulb of an air-thermometer T, whose tube is bent twice at right angles for the purpose. While the water is being heated by a lamp-flame, the gradual sinking of the fluid in the stem of the thermometer shows the rise of temperature; but, when the water fairly boils, the fluid stops sinking, showing that the temperature no longer rises. The fluid will remain motionless until the water in the vessel has been changed to steam. Let the bulb be lifted into the steam above the water; no change occurs in the height of the fluid in the stem, hence the temperature of the steam must be the same as that of the water.

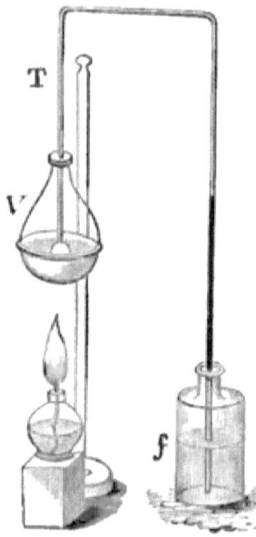

Fig. 66.

But the Expansion increases. — If all the heat applied to a boiling liquid is expended to produce expansion, we may expect that this effect will be more rapid than when a part of it was used to raise the temperature. This inference is abundantly verified. Steam fills about 1,700 times as much space as the water from which it was formed.

The Latent Heat of Steam. — The amount of heat required to expand water into steam, in other words, *the latent heat of steam*, is very great. By accurate experiment it has been found to be 972° F. The steam and the boiling water are of the same temperature, yet there is an excess of heat in an ounce of steam, which, if applied to one ounce of water, would be sufficient, were it possible, to raise its temperature to 972°; it will raise the temperature of *nine ounces* of water 108.°

60. The heat which has been required to produce expansion will be reproduced when the expanded body again contracts.

Heat restored after being used. — The following experiment very satisfactorily shows that heat is required to produce expansion, and that it is again given off when contraction occurs. The bulb of an air-thermometer is placed in a receiver over the plate of an air-pump; the stem, passing through the top of the receiver, is bent twice at right angles, and is filled with its colored fluid to a height very carefully marked. By a few rapid strokes of the piston, the air is partly exhausted from the receiver; that which remains *expands*, and the rising of the fluid in the thermometer shows that the bulb is at the same time *cooled*. If, now, air be allowed to return to the receiver, the air inside becomes *more dense*, and the sinking of the fluid in the thermometer shows that heat is again given off.

Illustrations. — Numerous familiar facts illustrate this principle. If water evaporates from the hand, it cools the hand, because the hand furnishes the heat to expand the water into steam; and, when vapor condenses upon the hand, the hand is warmed, because the vapor gives to it all the heat which had been used to keep the water in the form of steam.

Bodies in contact with melting snow or ice are cooled, because they must furnish heat to change the solid to the liquid form. But bodies near to freezing water are warmed, because the water, while freezing, gives up the heat which had been needed to keep it in a fluid form.

Steam-Heating. — In the production of steam the kinetic energy of the fuel, burning in the furnace, becomes potential energy in the expanded steam. Let this steam be condensed in pipes, and its potential energy is changed back again to kinetic energy, and the pipes show it by being hot. This principle is largely applied in the warming of buildings by steam.

REVIEW.

I.—SUMMARY OF PRINCIPLES.

Conductivity is that quality of a substance which allows heat to pass from molecule to molecule through the body.

Convection is the transmission of heat by moving particles or currents in the substance.

Solids are the best conductors, liquids stand next in order, and gases last.

Gases are heated by convection only, liquids by convection chiefly, and solids by convection never.

With few exceptions the volume of a body increases with the temperature.

Solids expand least, liquids more, and gases most, by the same increment of heat.

Thermometers are instruments for measuring temperature. $180° F. = 100° C. = 80° R.$

$Fahrenheit° = \frac{9}{5} Cent°. + 32, = \frac{9}{4} Réaumur° + 32.$

Sensible heat is the kinetic energy of the molecular motion. Latent heat is the potential energy due to the separation of molecules by heat.

Sensible heat becomes latent during the melting of a solid and the boiling of a liquid.

Latent heat becomes sensible during the liquefaction of vapors and the solidifying of liquids.

Equal weights of any two substances at the same temperature contain different quantities of heat.

The total heat in a body compared with that in the same weight of water at the same temperature is its specific heat.

The boiling point of a liquid depends on its purity, on the nature of the vessel containing it, and on the pressure upon it.

II. — SUMMARY OF TOPICS.

53. Conduction of heat. — Explanation. — Definitions. — Conductivity of metals. — Of liquids. — Of gases.

54. Convection illustrated. — Defined. — Of air. — Of liquids. — Of solids.

55. The twofold action of heat. — Expansion of solids. — Of liquids. — Of gases.

56. Temperature measured by expansion. — The thermometer. — Fahrenheit's. — Centigrade. — Réaumur's. — Degrees below zero. — Other forms.

57. Effect of heat on the motion and position of molecules. — Sensible and latent heat. — Specific heat.

58. The melting point. — No rise of temperature while melting. — But an increased expansion. — Except in the case of ice. — The latent heat of water.

59. The boiling point. — Depends on the purity of the liquid. — On the nature of the vessel. — And on the pressure. — No rise of temperature during ebullition. — But an increased expansion. — The latent heat of steam.

60. Heat restored after being used. — Illustrations. — Steam-heating.

CHAPTER VI.

ON UNDULATORY ENERGY: SOUND.

SECTION I.

ON THE TRANSMISSION AND REFLECTION OF SOUND.

61. Sound-waves travel through all elastic media. The Velocity of sound is not the same in different substances; it is governed by two Laws:—

1st, The velocity of sound varies inversely as the square root of the density of the substance.

2d, The velocity of sound varies directly as the square root of the elasticity of the substance.

In the same medium, the velocity of sound is uniform. (*G.* 226, 227; *A.* 489–491.)

Sound-Waves.—All sounds are undulations, but it is not true that all undulations are sound. Some are too slow to affect the ear: such are those produced by the vibrations of a cord not over-stretched. On the other hand, there are undulations which are too rapid to be heard. The lower limit has been fixed at sixteen a second, and the higher at 38,000. The energy of waves which occur within these limits of velocity can affect the ear, and such waves are called Sound-Waves, or simply, Sound.

It is interesting to notice that the limits of hearing are not the same in all persons. "Nothing can be more surprising than to see two persons, neither of them deaf, the one complaining of the penetrating shrillness of a sound, while the other maintains that there is no sound at all." "In the

'Glaciers of the Alps' I have referred to a case of short auditory range noticed by myself in crossing the Wengern Alp in company with a friend. The grass at each side of the path swarmed with insects which, to me, rent the air with their shrill chirruping. My friend heard nothing of this, the insect-music lying quite beyond his range of audition." (See Tyndall's Lectures on Sound.)

Are transmitted through all elastic Bodies. — Numerous facts, easily verified, prove that sound-waves can traverse all elastic bodies. When, for example, the blows of a hammer fall upon one end of a long wooden beam, an ear placed in contact with the other end hears the sound with surprising distinctness. The same thing is true of other solid bodies. The clatter of horses' hoofs, or the rattle of a railway-train, quite inaudible to one who stands erect, is heard distinctly when the ear is placed in contact with the ground. The solid earth, in this case, transmits the sound-waves.

In liquids, also, sound-waves travel freely. Let two stones be struck together under water; the sound will be heard by an ear, itself under water, a long distance away.

The transmission of sound-waves through gases is sufficiently familiar; the sounds which throng the ear so constantly are transmitted through the atmosphere.

The Velocity not the same in all Media. — The velocity of sound in a great many substances has been found by laborious and skillful experiments (see Tyndall's Lectures on Sound). In the following table some of these results are collected: —

Substances.	Temperature.	Velocity.
Air	32° F.	1,092 feet per sec.
Air	61° "	1,118 " " "
Oxygen	32° "	1,040 " " "
Hydrogen	32° "	4,164 " " "
River-water	59° "	4,714 " " "
Iron	68° "	16,822 " " "
Pine Wood	-	10,900 " " "

The velocity of sound depends upon the *density* and the *elasticity* of the medium in which it travels.

The first Law. — The density of oxygen, other things being equal, is about sixteen times that of hydrogen. But we see in the table that the velocity of sound in oxygen is only about one-fourth as great as in hydrogen. In this case *the velocity is inversely as the square root of the density of the medium*. It is always so. The law has been verified by repeated experiments.

The second Law. — When air is heated in a tight vessel, its elasticity is increased, while its density is unchanged. In this condition it will conduct sound more rapidly. If the elasticity of air be made four times as great, the velocity of sound will be doubled. *The velocity of sound* in this case *is directly as the square root of the elasticity of the medium*. It is so in all cases.

Density and Elasticity. — It is evident that both density and elasticity must be known, before we can judge the power of a substance to conduct sound. Liquids are, for example, more dense than gases: their conducting power, on this account, would be less; but, on the other hand, their elasticity *measured by the force required to compress them* is vastly greater, so that, as the table shows, water conducts sound better than air.

But in the same Medium Velocity is Uniform. — The velocity of sound-waves in air or in water, for example, is uniform. Moreover, all sounds in the same medium travel with the same velocity.[1] When we listen to the music of a distant band, the various notes, high and low, loud and soft, reach the ear in the same order in which they were made. So also the shrill chirping of insects, the dull thud of a falling stone, the melodious songs of the birds, and the murmur of rivulets, are all borne with equal swiftness through the air.

---

[1] There is reason to believe that very loud sounds do travel a very little faster than feeble ones.

Application. — So uniform is the velocity of sound, that distances may be measured by means of it. Suppose the flash of a cannon on a distant hill was seen, and in ten seconds afterward the report was heard, the temperature at the time being 61° F. The velocity of sound is 1,118 feet; and the sound-waves, starting when the flash was seen, took ten seconds to reach the ear. 1,118 × 10 = 11,180. The observer was at a distance of 11,180 feet from the cannon.

The Formula. — The two laws of velocity are expressed in the formula: —

$$V = \sqrt{\frac{e}{d}}.$$

The Effect of Temperature. — The heating of air expands it: its elasticity is thereby increased, and its density diminished. But to increase the value of e, and lessen that of d, would, in both cases, increase the value of the fraction $\frac{e}{d}$. Hence *a rise of temperature must increase the velocity of sound.*

This is verified by experiment. Notice the two velocities of air in the table.

62. When sound-waves fall upon the surface of a second medium, only a part of them enter: the rest are reflected. The Reflection of sound is governed by the following Law: —

The angle of reflection must be equal to the angle of incidence.

An Echo is produced by the reflection of sound. (*G.* 231, 232; *A.* 495, 496.)

The Reflection of Sound. — To illustrate the reflection of sound, suppose the line I A (Fig. 67) to represent the direction of several sound-waves, which, passing through the air, strike a body M M. Some of these waves will pass through the body, but others will be

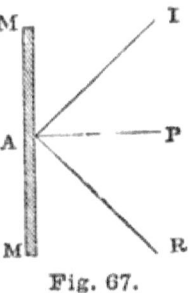

Fig. 67.

thrown off in the direction A R. These are the reflected waves.

Now, a person standing at R will hear the voice of another at I, when the distance is considerable, sounding as though it came from a person in the direction R A. We always judge the direction of a sounding body from us, to be that *from which the waves enter the ear.*

The Law of Reflection. — To understand the language of this law, let us refer again to Fig. 67. The waves I A, those which fall upon the reflecting surface, are called the INCIDENT waves: the waves A R, those that are thrown off from the surface, are called the REFLECTED waves; and the point A is called the POINT OF INCIDENCE. Now, if a perpendicular, A P, be drawn to the reflecting surface at the point of incidence, then the angle I A P is the ANGLE OF INCIDENCE, and the angle P A R is the ANGLE OF REFLECTION. The *law of reflection* requires that these angles shall always be equal.

The Echo. — An echo is a repetition of sound produced by the reflection of waves from a distant object. Who, after loudly uttering a word or sentence, has not sometimes listened to the sound of his own voice coming back to him from a distant wood, or from the face of a cliff, or, it may be, from the wall of a distant building? Visitors to Cooperstown, N.Y., will not soon forget the fine echo returned from the rocky hills which skirt Otsego Lake. There, we are told by Fenimore Cooper, once dwelt Natty Bumpo, the hero in the story of "The Pioneers." Let his name be loudly called from a certain place upon the lake, and immediately the response, "Nat-ty-Bum-po," every syllable full and clear, rings back over the water as if spoken by the hero himself from his cave in the cliffs.

Multiple Echo. — When two obstacles are opposite to one another, the sound may be reflected back and forth many times. Surprising repetitions of echoes are in this way sometimes produced. It is said that an echo near Milan repeats a single sound thirty times. "When a trumpet is sounded at the proper place in the Gap of Dunloe, the sonorous waves reach the ear after one, two, three, or more reflections from the adjacent cliffs, and thus die away in the sweetest cadences."

SECTION II.

ON MUSICAL SOUNDS.

63. Musical sounds are caused by rapid vibrations which follow each other with great regularity. Any noise whatever, when repeated rapidly, will cause a continuous tone: even separate puffs of air, following each other rapidly, produce a musical sound. (G. 236, 237; A. 512.)

Musical Sounds. — When a single and intense air-wave is suddenly produced, as when a gun is fired, the resulting sound is called a REPORT. Let a series of such sounds be made in quick but irregular succession, and the resulting sound is called NOISE. But when the waves are made with regularity, and follow each other so swiftly that the ear can distinguish no interval of time between them, the result is a MUSICAL SOUND.

Any Noise repeated rapidly causes a continuous Tone. — No matter what the source of the waves may be, nor how unmusical the separate noises: only let them be repeated with regularity and rapidity, and they will result in music. Slowly pass a piece of ivory, or even the fingernail, over the rough surface of a wound piano-wire, and the sound of its strokes against the separate ridges is altogether unpleasant; but pass it *quickly* over the same surface, and the ear is saluted with a musical tone of surprising shrillness and purity. If a card be pressed against the teeth of a wheel which rotates slowly, a series of distinct and unpleasant taps will be heard; but if, by means of a larger wheel and band, this wheel be made to revolve rapidly, the taps will coalesce, and salute the ear with music.

Even Puffs of Air: the Siren. — The *siren* is an instrument by which a series of air-puffs are made to produce a musical sound, and by which the number of puffs made in a second are registered. Its structure may be learned from Fig. 68.

Description. — A brass tube, O, leads from a wind-chest, E, to a brass plate, B, which is pierced with a series of holes arranged around the circumference of a circle. Above this plate is a disk, also perforated with holes exactly corresponding to those in the plate below. The disk is provided with a vertical steel axis, and is so fixed that it may rotate with a very small amount of friction. The wheel-work shown in the upper part of the figure registers the number of puffs made in any given time.

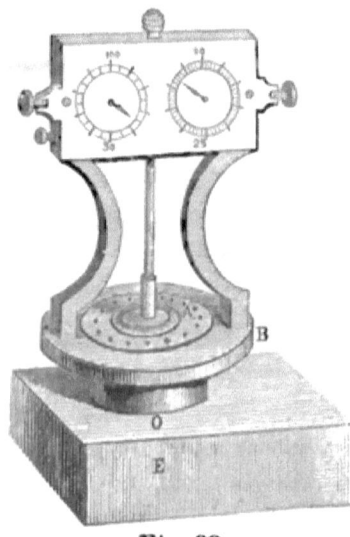

Fig. 68.

Its Action. — Now, when the disk revolves, the holes in it will be brought alternately over the perforations in the plate B, and the spaces between them, so that these holes will be alternately opened and closed. When the disk is still, and the holes are open, if air be urged through the tube O, it will escape from the top in steady streams, but when the disk revolves these streams will be cut up into successive puffs. If the disk turns slowly, the separate puffs are heard; but, as the disk is turned more and more rapidly, the air announces its escape by a musical sound of great purity and increasing shrillness.

By a simple artifice, the air which gives the sound is made to turn the disk. This is done by making the holes through the plate B oblique instead of vertical; those in the disk being also oblique, but inclined in the opposite direction.

64. Musical sounds differ in three respects: in Pitch; in Intensity; and, in Quality.

I. — PITCH.

Pitch depends entirely upon the rapidity of vibrations which produce the sound.

The difference in the pitch of two sounds is called an Interval; and a series of eight sounds of different pitch has been adopted as the foundation of all music, and called the Diatonic Scale.

The Number of vibrations to produce the note A of the treble clef is 440 a second. (*G.* 243-248; *A.* 520.)

Pitch depends on the Rapidity of Vibration. — The pitch of sounds is that which distinguishes them as being high or low. It depends entirely upon the rapidity of vibration: the more rapid the vibrations, the higher will be the sound produced. Two sounds made by the same number of vibrations per second, however much they may differ in other respects, will have the same pitch.

Intervals. — When the number of vibrations which produce one sound is twice as great as that which produces another, we must not say that the sound is *twice as high*, but rather that it is an *octave above*. The term "octave" is used to designate a tone which is made by twice the number of vibrations needed to produce a lower one, called the fundamental. Other intervals will be named in the description of the scale.

The Diatonic Scale. — Now, the difference in pitch, or the interval between a fundamental note and its octave, is very great. To fill up this interval, sounds have been chosen which blend, or harmonize most perfectly, with the fundamental, or with each other. These, placed between the fundamental and its octave, form a series of eight tones, called the Natural or the Diatonic Scale.

The eight notes of the scale are expressed by the following names and intervals: —

Names,   C, D, E, F, G, A, B, C.
Intervals, 1st, 2d, 3d, 4th, 5th, 6th, 7th, 8th.

This scale repeated about eleven times, making what is termed in music eleven octaves, will include all sounds within the range of the human ear. Only about seven octaves are used in music.

The method of representing the notes in music is familiar to all. Remember that the note called A *is found in the second space of the treble clef*, and the position of all others may be easily traced.

Number of Vibrations for the Notes. — The number of vibrations to produce the various notes may be found by experiment with the siren. (See Tyndall on Sound.) Or it may be found with the greatest accuracy by means of the electric register (Fig. 52). Fig. 69 shows how the standard tuning-fork is made to open and close the electric circuit.[1]

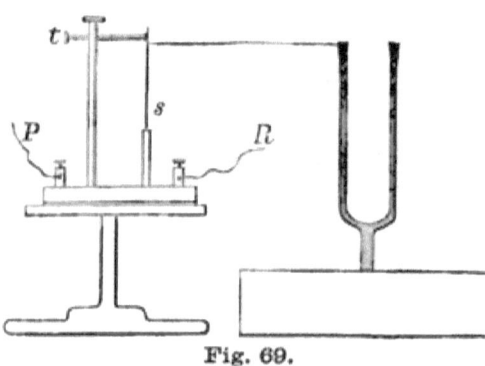

Fig. 69.

---

[1] From the prong of the fork a silk fiber stretches across to a slender vertical spring, *s*, fixed at the lower end, while its upper end rests against the end of a set screw, *t*. The two surfaces in contact are platinized. The spring is slightly bent; and the set screw, pressing against its upper end, holds it in constant tension, and thus forbids it vibrating, except in unison with the fork. Every vibration of the prong is transmitted by the fiber, and compels the spring away from contact with the screw. Putting this break-circuit in place of the monocord, represented in Fig. 52, between the pendulum *P* and the register *R*, the vibrations of the fork record themselves upon the moving paper.

Noticing that the accuracy of Koenig's tuning-forks is questioned by Mr. Ellis (*Nature*, xvi., p. 85), I fancied that the testimony of this method would not be without interest. Seizing the earliest opportunity, therefore, I submitted the *Ut₃* fork, bearing Koenig's monogram, to careful examination. The pendulum was accurately adjusted to hold the circuit one-half a second. The iodide-starch solution, with a battery of ten Bunsen's immersion cells, was used. Fifteen perfectly distinct and easily counted records were taken. Every one of these autographs was found to consist of 128 dots, representing 128 complete vibrations per half-second, or 256 per second, testifying to the exactness of Koenig's stamp. *Jour. Frank. Ind.* vol. lxxii.

On the English standard of pitch, the note A of the treble clef is made by 440 *vibrations a second*.

On the French standard the rate for the same note is 435.

Relative Rates. — If we represent the number of vibrations for the fundamental note by 1, then the several notes of the scale will be made by the following ratios: —

C, D, E, F, G, A, B, C.

1, $\frac{9}{8}$, $\frac{5}{4}$, $\frac{4}{3}$, $\frac{3}{2}$, $\frac{5}{3}$, $\frac{15}{8}$, 2.

Absolute Rates. — Now, remembering this series of fractions, and the fact that A is made by 440, the number of vibrations for all the others may be found. Thus, for example, how many vibrations to give the fundamental C? The *relative* number of vibrations for A and C are $\frac{5}{3}$ and 1; that is, A is produced by $\frac{5}{3}$ as many vibrations as C; or, to reverse the ratio, C requires $\frac{3}{5}$ as many as A, and $\frac{3}{5} \times 440 =$ 264. Having this number for the fundamental, multiply this by the fractions $\frac{9}{8}$, $\frac{5}{4}$, $\frac{4}{3}$, &c., and the numbers for the corresponding notes will be obtained. These multiplied by 2 will give the numbers for the notes in the next higher octave; or, divided by 2, will give the numbers for the notes in the octave below.

II. — INTENSITY.

Definition. — The intensity of sound is that which distinguishes it as being loud or soft. It is the energy of the waves against the ear. It depends *entirely* upon the *amplitude* of the vibrations which produce it. The greater the amplitude, the louder the sound will be. In the case of a vibrating string, for example, the *loudness* or intensity of the sound made by it will depend entirely upon the distance through which the string vibrates across its line of rest. (*G*. 222, 223.)

The Law. — If this distance be doubled, then the velocity of a string will be doubled also. But a body with double *velocity* has a quadruple *energy*. Hence the intensity of sound varies as the square of the amplitude.

III. — QUALITY.

Definition. — By quality, we refer to that peculiarity of sound by which we may distinguish notes of the same pitch and intensity, made on different instruments. The pitch and intensity of notes made on a violin and on a piano may not differ, and yet how easy to tell the sounds apart! We recognize the voices of friends, not by their pitch nor their intensity, but by their *quality*. (G. 249, 253–255; A. 523–526.)

Compound Sounds. — A musical sound is rarely made by a single undulation: several, with different rates and amplitudes, are combined in almost every tone we hear.

Fig. 70.

Illustration. — If a plate of brass is fastened at its center, and covered with a sprinkling of fine sand, and then if a violin-bow is drawn across its edge (Fig. 70), a shrill tone is produced. The sand at the same time dances about in a curious manner, and finally arranges itself in straight lines and curves upon the plate.

The sand gathers *where there is least vibration*, and shows that the plate vibrates in *segments* (p. 104). The plate in the cut shows eight segments.

Now, each of these vibrating parts starts an undulation in the air, and consequently there are eight different sounds combined in the one shrill tone which is heard.

Overtones. — Whenever a piano-wire or the cord of a harp is struck, it vibrates as a whole and in segments at the same time.

Its vibration as a whole yields its *fundamental* tone.

Its *vibrations in segments* yield higher tones, called OVERTONES, or sometimes HARMONICS.

The fundamental with its overtones are combined in the sound which is heard whenever the wire or cord is struck.

Helmholtz' Theory. — The quality of a sound depends on the number and prominence of the different undulations which are combined to produce it.

Each different set of undulations produces a resultant wave of a different *form*, and these waves of different form affect the ear as sounds of different quality.

65. The Phonograph is an instrument for recording and reproducing sounds. It acts on the principle that waves of the same rate, amplitude, and form are exactly the same sound.

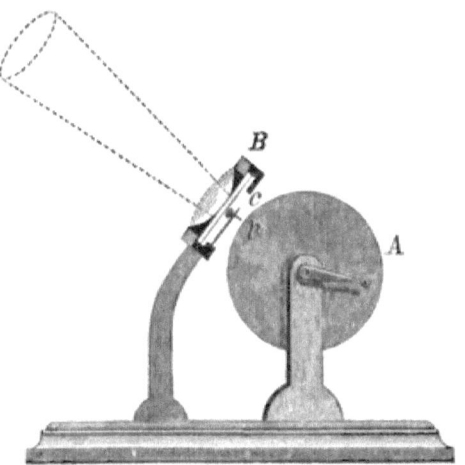

Fig. 71.

The Phonograph. — The construction of Edison's phonograph or "talking-machine" is shown in Figs. 71 and 72.

A spiral groove is cut around a cylinder of metal, A, which may be turned with a crank. On one end of its axle a screw-thread is cut, so that when the cylinder is turned it is at the same time moved forward. In front of this cylinder is a mouthpiece, B, with a thin disk of metal, C, which the voice will vibrate, and between the disk and the cylinder is a needle-point p which vibrates with the disk.

To record the Voice. — A sheet of tinfoil is wrapped smoothly around the cylinder, and the mouthpiece is then adjusted so that the needle-point will press lightly against its surface just over the spiral groove. The speaker then talks loudly into the mouthpiece while the cylinder is turned. The air-waves of the voice vibrate the disk; and the vibrating disk drives the needle-point against the tinfoil, and presses it to greater and less depths into the groove. This tracing of the needle is a record of the voice.

The *closeness* of the depressions represents the pitch.
The *depth* of the depressions represents the loudness.
The *form* of the depressions represents the quality.

To reproduce the Voice. — Move the mouthpiece so that the point no longer touches the tinfoil, and turn the cyl-

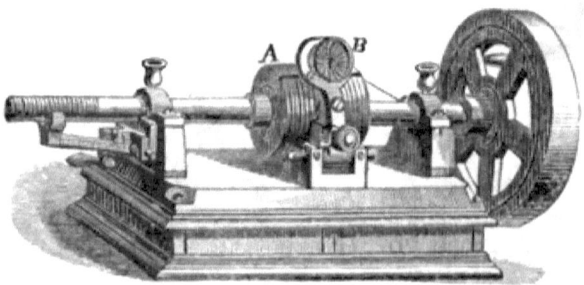

Fig. 72.

inder back to the place it was in when the record began. Then again bring the point in contact with the tinfoil and turn the cylinder forward exactly as at first. The needle-point will follow the tracing which it made before, and in going over the prominences and into the depressions will compel the disk to repeat the same vibrations which it made before under the influence of the voice. These vibrations of the disk reproduce the air-waves which the voice threw against it, and if the ear be placed near the mouthpiece, or, better still, if a large cone, shown by dotted lines in Fig. 71, be used, the utterances of the speaker will be repeated by the phonograph.

66. Musical instruments are, for the most part, of two classes: first, those in which the sounds are produced by vibrating strings, and, second, those in which sounds are made by vibrating columns of air.

I. — STRINGED INSTRUMENTS.

Stringed Instruments. — The violin, the guitar, and the piano are familiar forms of stringed instruments. In every case, cords or wires are tightly stretched over some solid body having considerable surface. The music of these instruments is not made by the vibrations of their cords alone: the simple vibration of a cord is not able to produce sound of sufficient intensity; but by being stretched over hollow boxes made of elastic wood, the material of the box, and the air inside of it, are made to vibrate, and these vibrations, joined with those of the cords, produce the sounds of the instrument.

Pitch varied by using Strings of different Lengths. — The pitch of any sound depends upon the rapidity of vibrations; but according to the first law of vibrating strings, the rapidity of vibration is greater as the string is made shorter. To obtain sounds of different pitch, we may, then, use strings of different lengths.

Illustration. — Suppose we wish to know the lengths of eight strings of the same weight and tension, which will give the eight notes of the scale. We have learned that the number of vibrations per second is *inversely* as the length of the cord, and we have learned also that the relative numbers of vibrations for the eight notes are expressed by the series $1, \frac{9}{8}, \frac{5}{4}, \frac{4}{3}, \frac{3}{2}, \frac{5}{3}, \frac{15}{8}, 2$. Then *invert* the terms of this series, and they must express the relative lengths of cord to produce the notes. They will be $1, \frac{8}{9}, \frac{4}{5}, \frac{3}{4}, \frac{2}{3}, \frac{3}{5}, \frac{8}{15}, \frac{1}{2}$. Knowing the length of the string to give the fundamental, it is easy to calculate the lengths of all the others. Let us start with a string 18 inches long for the first note; the second must be $\frac{8}{9} \times 18$; the third must be $\frac{4}{5} \times 18$; the fourth must be $\frac{3}{4} \times 18$; and so on until the eighth, which must be $\frac{1}{2} \times 18$.

Pitch is varied by using Strings of different Tension. — According to the second law of vibrating strings, the number of vibrations made in a second increases when the tension increases. Hence the pitch of sound made by the string will be higher when the tension is made greater.

Pitch is varied by using Strings of different Weights. — According to the third law of vibrating strings, the number of vibrations made in one second varies inversely as the square root of the weight of the strings. Hence the pitch of the sound will be higher, when the string which makes it is lighter.

II. — WIND-INSTRUMENTS.

Wind-Instruments. — The organ and the clarionet are examples of wind-instruments. In the organ, sounds are made by vibrating columns of air in pipes, sometimes aided by the vibrations of a slender and elastic tongue, called a reed. (See Tyndall on Sound.) In the clarionet the sounds are always made by air-vibrations aided by a reed.

Organ-Pipes. — The pitch of sounds in pipes depends upon the lengths of the pipes. A pipe, to produce the *lowest* note used in music, must be thirty-two feet in length, and the pitch of tones from other pipes will vary inversely as the lengths of the pipes.

Organ-pipes are sometimes open at the top, and sometimes closed. An open organ-pipe yields a note an octave higher than a closed pipe of the same length. A closed pipe, to give the lowest note in music, need only be sixteen feet in length.

SECTION III.

ON MUSICAL AND SENSITIVE FLAMES.

67. When a gas-flame burns within a glass tube, a musical sound is produced. The pitch of the tone depends on the length of the tube and the size of the flame. A silent flame may be made to sing by sounding near it the note of the tube. Naked flames are also sensitive to the action of neighboring sounds. (*G.* 281.)

The Musical Flame. — Let a flame of common coal-gas, placed under the end of a glass tube T (Fig. 73), be slowly raised into it; when a particular height is reached, the flame, if small enough, will burst forth into a loud and continuous sound. This sound is often harsh, sometimes melodious.

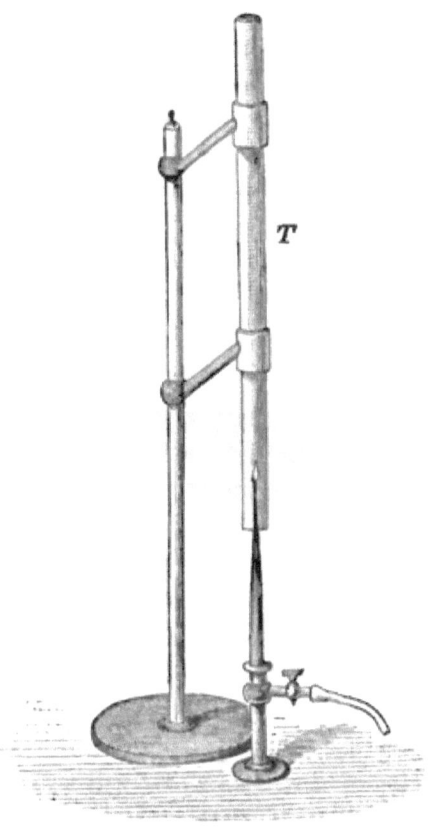

Fig. 73.

At the beginning it is sometimes low and smooth, like the whistle of a very distant locomotive; but, as the experiment goes on, the intensity of the sound rapidly increases until, like the long monotonous screech of the engine at hand, it becomes almost unbearable.

In tubes of tin and pasteboard, sounds of different quality are obtained.

Explanation. — That a gas-flame flutters when exposed to a gentle breeze, is a fact sufficiently familiar. Now, this fluttering flame, like a vibrating tongue or reed, must cause vibrations in the air around it. This is the key to the explanation of musical flames.

The air in the tube is heated by the flame; it rises, and an upward current through the tube is thus produced. The flame flutters in this current, and causes a system of waves, whose rapidity and amplitude give pitch and intensity to the note produced. If we inquire further about the cause of the fluttering, we are told that experiments by Faraday proved that gas issues from a burner in an *unsteady* stream, due to the friction against the sides of the tube, and that in burning it makes a series of inaudible explosions. A current of air heightens both of these effects, and makes them sensible.

Evidence that the Flame is intermittent. — That a musical flame is thus intermittent, is shown by the following beautiful experiment. The tube T (Fig. 74) is blackened so as to keep the light from falling on the screen placed behind it at S. A concave mirror M, in front of the flame, forms an inverted image of it on the screen. If the mirror is turned horizontally, while the flame is silent and steady, the image will move, and if the motion of the mirror is swift, an unbroken band of light will be seen on the screen. But if, when the flame is singing, the mirror is swiftly turned, a series of *distinct images* will appear.

The Pitch of the Note. — In these tubes, as in organ-pipes, the pitch of the sound depends upon the length of the tube. But, while the pitch depends chiefly upon the length of the tube, it is partly governed by the size of the flame. If one tube is just twice the length of another, its fundamental note is an octave below; but, when placed over a flame whose size fits it to sing in the shorter tube, the note of the shorter tube will be produced. Then let the flame be gradually enlarged, and in a little time the low fundamental note of the tube suddenly bursts forth. By varying

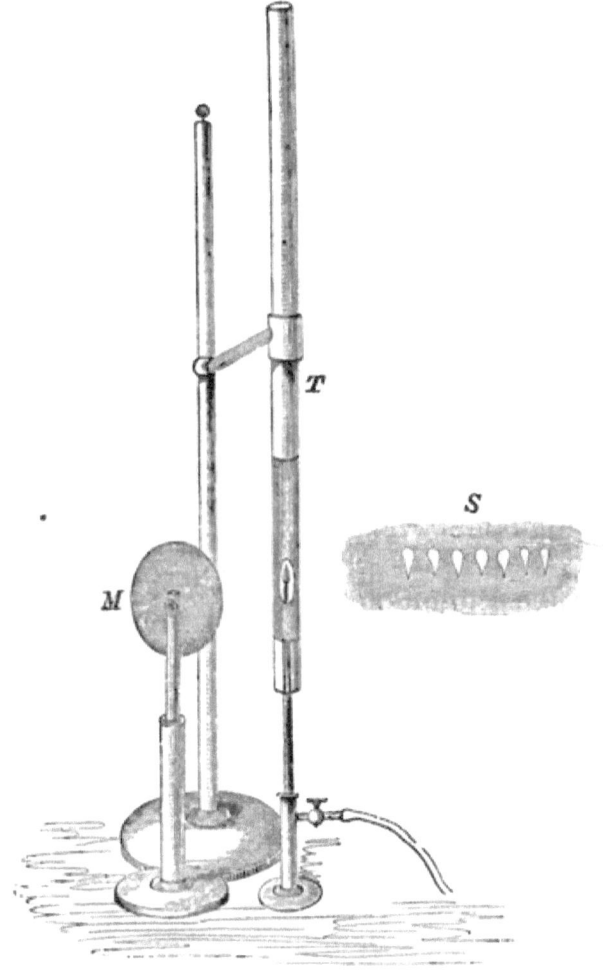

Fig. 74.

the size of the flame it is possible to obtain the fundamental note, its octave, and its four harmonics, from the same tube.

Sensitive Flames. — The name "sensitive flames" is given to those which, without being inclosed in tubes, are affected by sounds. Certain sounds in an instrumental concert often cause curious motions of the gas-flames in the room. This observation was first published in 1858. The motions referred to consist of a "jumping" of the flame to considerable height, or a thrusting forth of tongues of flame from its upper edge. (See Tyndall on Sound.)

An effect just opposite this, the shortening of tall flames, was first noticed by Mr. Barrett of London, in 1865. He says ("Chemical News, American Reprint," July, 1868), "A jet of gas issuing from a V-shaped orifice was found to be quite insensible to sound until the flame reached a height of ten or twelve inches (see Fig. 75); and then, at the sound of certain high tones, the flame shortened, and spread out into a fan-shape." Another flame he thus describes: "So sensitive is this flame, that even a chirp made at the far end of the room brings it down more than a foot. Like a living being, the flame trembles and cowers down at a hiss; it crouches and shivers as if in agony at the crisping of this metal foil, though the sound is so faint as scarcely to be heard; it dances in tune to the waltz played by this musical box; and, finally, it beats time to the ticking of my watch."

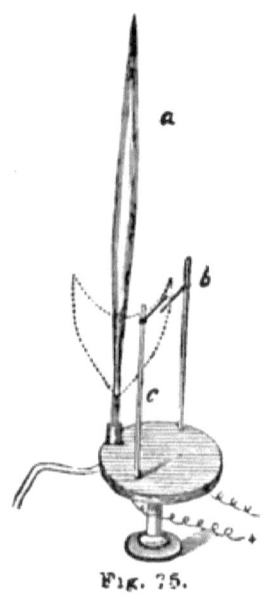

Fig. 75.

Mr. Barrett also suggests that these flames may yet be turned to some use, and, to illustrate, suggests an arrangement shown in Fig. 75.

Near the tall sensitive flame a, stand two vertical brass rods, $b\ c$. Projecting from these rods are two metallic ribbons, made of layers of silver, gold, and platinum welded together. The ends of the ribbons are about half an inch apart. By heat the different metals expand unequally, and, bending the ribbons, bring their ends together. The brass rods are connected by wires with an electric bell at a distance.

Now, as long as the tall flame is not disturbed, the metallic ribbons are not in contact: the circuit is broken, and the bell is silent; but at the sound of a whistle the flame jumps down, warms the ribbons, completes the electric circuit, and rings the distant bell.

SECTION IV.

REVIEW.

I. — SUMMARY OF PRINCIPLES.

Undulations with rates between 16 per second and 38,000 per second will affect the ear as sound.

Such undulations can traverse all kinds of elastic bodies, but not with the same velocity.

The velocity in a denser medium is less than in a rarer one. In a more elastic medium it is greater than in one less elastic.

The laws are expressed by the formula: —

$$V = \sqrt{\frac{e}{d}}.$$

When sound-waves flow against an obstacle, they are thrown back, or reflected. The angle of reflection equals the angle of incidence.

Any sound repeated with great regularity and rapidity becomes a musical tone.

Tones differ in pitch, in intensity, and in quality, or, as the French call it, *timbre*.

The pitch depends upon the *rate* of the waves.

The intensity depends upon the *amplitude* of the waves.

The quality depends upon the *form* of the waves.

Waves of the same rate, amplitude, and form will be recognized as the same sound, in whatever way they may be produced.

II. — SUMMARY OF TOPICS.

61. Sound-waves. — Traverse elastic bodies. — With different velocity. — The first law. — The second law. — Velocity uniform in the same medium. — Application. — The formula. — Effect of temperature.

62. Reflection of sound. — The law. — Explanation of the echo. — Multiple echo.

63. Musical sound. — Continuous tone by rapidly repeated impulses. — The siren. — Description of the siren. — Explanation of its action.

64. Pitch depends on rate. — Intervals. — The diatonic scale. — English and French standards of pitch. — Relative rates for the eight notes. — Absolute rates. — Intensity. — Quality defined. — Compound tones. — Illustration. — Overtones. — Helmholtz' theory of quality.

65. The phonograph. — To record the voice. — To reproduce the voice.

66. Stringed instruments. — Pitch varied by strings of different lengths. — Different tensions. — Different weights. — Wind-instruments.

67. The musical flame. — Pitch of the tone. — The sensitive flame.

CHAPTER VII.

ON RADIANT ENERGY: LIGHT.

SECTION I.

ON TRANSMISSION.

68. RAYS of light are transmitted through some media more freely than through others, but always according to two laws: —

1st, In a medium of uniform density, light goes in straight lines with a uniform velocity.

2d, The intensity of light varies inversely as the square of the distance from its source.

The art of Photometry depends upon this second law. (*G*. 489, 491–493, 495; *A*. 795, 803.)

Wave-Front. — When a pebble falls into quiet water, a wave advances outward like the rim of an expanding wheel. The *outside* of this circular wave is the wave-front.

When a gas-jet is lighted, a wave darts outward, advancing in every direction like the surface of an expanding sphere. The outside of this spherical wave is the wave-front.

The outside of an advancing wave of any kind is called the WAVE-FRONT.

Rays of Light. — A single line of light, or, more accurately, the path of a single point in the wave-front, is called a RAY of light. But a ray of light is quite too delicate a thing to be seen. The smallest portion of light which can be separated by experiment consists of many rays. A collection of parallel rays is called a BEAM of light. A collection of rays which diverge from a point, or which converge toward a point, is called a PENCIL of light.

Rays of Light are transmitted.—Some substances permit light to pass through them freely; they are said to be *transparent*. Air and water are examples of transparent bodies. Others, such as iron and wood, appear to forbid the passage of light through them: they are said to be *opaque*. But no substance will transmit all the light which it receives; even the air is not perfectly transparent. On the other hand, no substance will stop all the light which falls upon it; even gold, when a very thin leaf of it is examined, can be seen to transmit light. All substances are doubtless able to transmit light in some degree.

Light moves in straight Lines.—That light moves in straight lines, is shown by numerous familiar facts. We can not see through a crooked tube, simply because light can not pursue a crooked path. And again: who has not seen the sunlight coming through the shutters of a half-darkened parlor, spotting the opposite wall with circles of light? The sun, the hole in the shutter, and the spot on the wall, are always in the same straight line. Let the air of the room be sprinkled with dust, and the paths of the sunbeams are seen streaking the air with bars of light.

With uniform Velocity.—Light travels through space with a uniform velocity of about 186,000 miles a second. This number has been found by observing the eclipses of one of the moons of Jupiter. The time when the eclipse should begin can be calculated by an astronomer with great accuracy. But it is found that when the earth is in that part of its orbit nearest to Jupiter, the eclipse begins 16 minutes and 36 seconds sooner than it appears to when the earth is in the opposite part of its orbit. It must, therefore, take light 16 minutes and 36 seconds to go across the earth's orbit. When this distance is known, and divided by the number of seconds, the velocity of light is found. The result is about 186,000 miles a second. For all distances on the surface of the earth, the passage of light may be considered instantaneous. It would go around the world almost seven times in a single second.

The Second Law. — That the intensity of light varies inversely as the square of the distance, may be easily proved by experiment.

A square piece of stiff cardboard, A (Fig. 76), is placed in front of another, B, very much larger. If now a candle-flame be placed in front of the small card, a shadow will be cast upon the large one. This shadow will be larger as the small card is moved nearer to the flame; it will be smaller as it is moved the other way. The figure

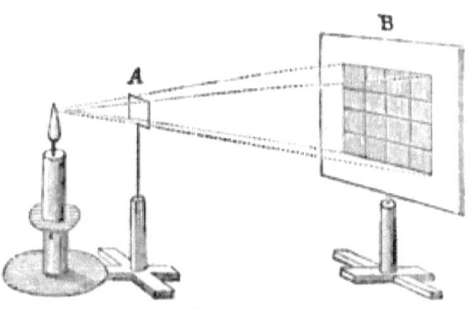

Fig. 76.

is intended to show the small card to be just one-fourth as far from the flame as the large one. In this case the shadow will be found to be exactly sixteen times as large as the card in front of it. Now, the same amount of light which is spread over the small card would, if it could go on, just cover the place of this shadow. But, if the same amount of light is spread over sixteen times as much surface in one case as in another, it can be only one-sixteenth as intense. At four times the distance from the luminous body, in this case, the intensity of the light is one-sixteenth as great. At three times the distance, the light would, in the same way, be found to be one-ninth as intense. In other words: *the intensity of light varies inversely as the square of the distance from the luminous body.*

Photometry. — It is often desirable to compare the *illuminating powers* of different flames. The art of doing this is called PHOTOMETRY. The simplest method is to place the two flames at such distances from a screen, that the intensities of the light they shed upon it shall be equal; the illuminating powers of the flames must then be as the squares of these distances. Suppose, for example, that we wish to know how many times more light one candle will give than another of inferior quality. Let a slender rod B (Fig. 77) be put just in front of a white screen A, and then move the flames to such distances that the two shadows of the rod, falling side by side

Fig. 77.

upon the screen, shall appear to be of *equal darkness*. The intensities of their lights on the screen must then be equal. Measure the distances from the flames to the screen: the amounts of light they give will be as the squares of their distances. One being twice as far away as the other, it gives four times as much light.

SECTION II.

ON REFLECTION.

69. When light in passing through one medium comes against the surface of another, only a part will enter; another part will be reflected, obeying the following law: —

The angles of incidence and reflection must be equal, and in the same plane. (*G.* 497; *A.* 881, 882.)

Reflection. — The reflection of light is in all respects like the reflection of sound.

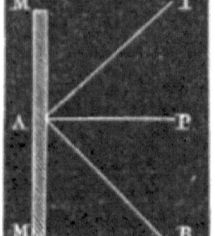

Fig. 78.

The same terms are used to describe it; the same figure may be reproduced to illustrate it. Thus in

Fig. 78, the line I A may represent a beam of light passing through air, and striking upon the surface of a plate of glass at A. One part of the beam will enter the glass, and emerge again on the other side; but another part will be thrown back into the air in the direction A R. The beam I A is the incident beam. The beam A R is the reflected beam. The point A is the point of incidence.

The Law of Reflection. — The reflection of light is also governed by the same law as the reflection of sound. The angle I A P (Fig. 78) is the angle of incidence. The angle P A R is the angle of reflection. These two angles must be equal.

Illustrations. — How various and beautiful are the phenomena which this principle of reflection explains! The sky, with all its floating clouds or shining stars, is painted in every pool of water, because the light from them, falling on the surface of the water, is reflected to our eyes. Rocks, and shrubbery, and dwellings along the shore, are pictured in the quiet waters of the lake, with skill exceeding that of any human artist.

Vision is produced by reflected light. How seldom do we receive the direct rays of the sun into the eye! How rarely, indeed, do we look directly upon any luminous body! But in all other cases we see objects only by reflected light. The sunbeams fall upon all objects exposed to them, and, bounding from their surfaces, enter the eye; and we see them in the direction from which the reflected rays have come.

70. The effects of mirrors are explained by reference to the law of reflection.

Rays of light reflected by a plane mirror have the same relation to each other as before reflection;

But the effect of a concave mirror is to collect the rays of light which are reflected by it;

While a convex mirror always separates the rays which it reflects. (G. 506.)

Mirrors. — Any surface smoothly polished, that will reflect nearly all the light which falls upon it, is called a MIRROR. The smooth surface of quiet water is a very perfect mirror. Artificial mirrors are generally made of metal or of glass. If made of glass, a thin film of mercury is spread over one side, and the smooth surface of this metallic coating is really the reflecting surface. Mirrors are either *plane* or *curved.* Of the curved mirrors there are two varieties, — the *concave* and the *convex.*

The effect of Plane Mirrors. — The rays of light which fall upon a mirror may be parallel, or converging, or diverging, but can have no other relation. Now, let the mirror be represented by the straight line A B (Fig. 79), and suppose, *first,* that it receive two parallel rays represented by the lines $a\,c$ and $b\,d$. At the point of incidence, c, erect a perpendicular to the surface A B. The angle $a\,c\,g$ will be the angle of incidence. Then draw the line $c\,f$. so as to make the angle of reflection, $g\,c\,f$, equal to the angle of incidence, and $c\,f$ must be the direction of the ray reflected from the point c. Again: at the point of incidence, d, erect a perpendicular, and draw the line $d\,e$, making the angle of reflection equal to the angle of incidence, and this line must represent the ray reflected from the point d. It will be found that the reflected rays, $c\,f$ and $d\,e$, will be parallel.

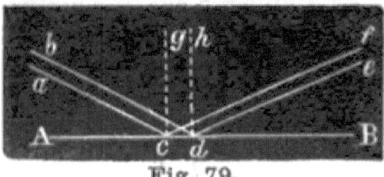
Fig. 79.

Suppose, *second,* that two rays, $a\,c$ and $b\,d$ (Fig. 80), are *converging,* and strike the mirror at the points c and d.

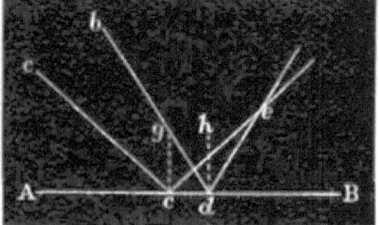
Fig. 80.

By making the angles of incidence and reflection equal, exactly as it was done in the preceding case, we find that the reflected rays will take the directions $c\,e$ and $d\,e$, *converging* to the point e.

Suppose, *third*, that the rays are *diverging*. Represent them by lines *a c* and *a d* (Fig. 81). Erect the perpendiculars, and construct the angles of incidence and reflection equal, and the directions of the reflected rays will be *c e* and *d f*, *diverging* from each other.

In each of these three cases, the reflected rays have the same relation as the incident rays.

The effect of Concave Mirrors. — We will notice only those concave mirrors whose surfaces are spherical. If we know the direction of the incident rays, we can find the direction of the reflected rays by making the angle of reflection equal to the angle of incidence.

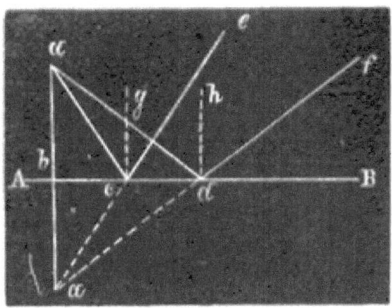

Fig. 81.

To construct the angle of incidence, we must, as in the plane mirror, erect a perpendicular to the concave surface at the point of incidence; and all difficulty disappears when we remember that *a perpendicular to any spherical surface is the radius of the sphere.*

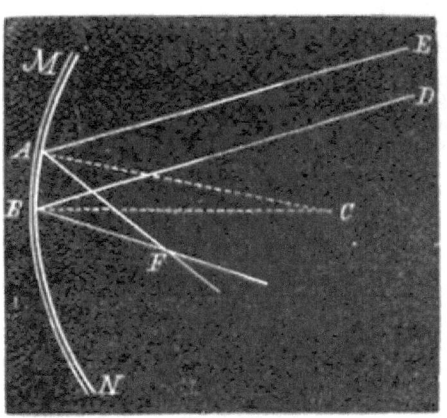

Fig. 82.

In Fig. 82, M N represents a section of a concave mirror. The point C represents the *center of curvature*, that is, the center of the hollow sphere of whose concave surface the mirror is a part. Now, if E A and D B represent two *parallel* incident rays, and we wish to find the direction they take after reflection, we may draw the radii C A and C B, making the angles of incidence E A C and D B C, and then draw the lines A F and B F, so as to make the angles of reflection equal to these. By so doing, we find that the reflected rays *converge*, and cross each other at the point F.

In Fig. 83, the lines E A and D B represent *converging*

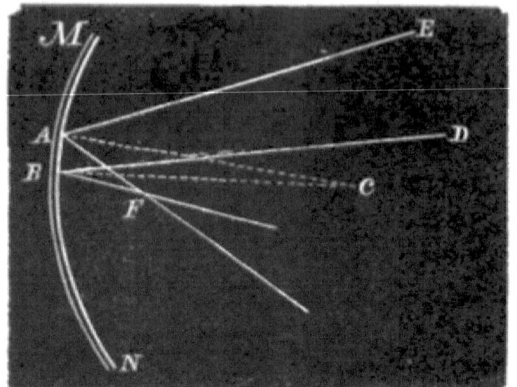

Fig. 83.

rays. By constructing the angles of incidence and reflection in the same way as before, we find that the reflected rays cross each other at the point F, *converging faster* after reflection than before.

In the same figure, F A and F B may represent *diverging rays*, striking the mirror at the points A and B. By constructing the angles of incidence and reflection equal, we find the reflected rays taking the directions A E and B D, *diverging less* after reflection than before.

Now, since parallel rays are made converging, and converging rays are made more converging, while diverging rays are made to diverge less, we may say that *the general effect of a concave mirror is to collect rays of light.*

Foci. — A *focus* is any point where rays of light cross, or appear to cross, after reflection. The points F, in Figs. 82 and 83, are foci. The *axis* of a mirror is a straight line drawn through the center of curvature and the middle point of the mirror.

In Fig. 84, the line C A is the axis of the mirror M N, whose center of curvature is at C. The focus of rays that are parallel to the axis, and fall upon the mirror near its middle point, is called the PRINCIPAL FOCUS. If the rays B E and D H (Fig. 84) are near and parallel to the axis C A, they will, after reflection, cross each other at the point

F, and this point is the *principal focus* of the mirror. *The principal focus is on the axis, half way between the center of curvature and the mirror.*

The effect of Convex Mirrors. — In Fig. 85, a convex mirror is represented by M N, its center of curvature by the point C. Two parallel rays of light, E A and D B, strike the mirror at the points A and B.

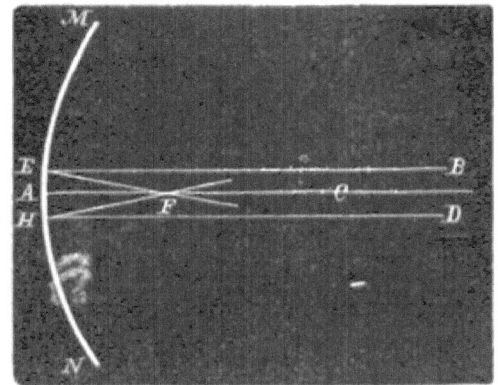

Fig. 84.

To construct the angles of incidence, we must erect perpendiculars to the surface at these points. The perpendiculars are the radii, C A and C B, *extended* beyond the convex

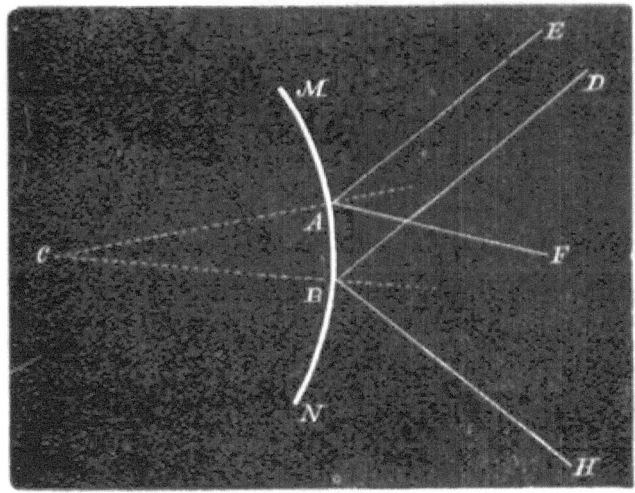

Fig. 85.

surface of the mirror. By making the angles of reflection equal to the angles of incidence, the reflected rays are found to take the directions A F and B H. We notice that parallel rays are rendered diverging.

So we might show that diverging rays would be made more diverging, and that converging rays would be made to converge less. We say, therefore, that *the general effect of a convex mirror is to separate rays of light.*

71. When the light reflected from a mirror enters the eye, we see an image of the object from which the light proceeds.

The image of any point will always be found where the rays of light which go from that point either meet, or appear to meet, after reflection.

The image formed by a plane mirror is always as far behind the mirror as the object is in front of it; the same size as the object, and erect. (*G.* 501–503; *A.* 883–885.)

Images by Reflection. — When an object, a lighted candle for example, stands before a looking-glass (Fig. 86), numberless rays of light from every point of it fall upon the mirror. These rays are reflected, and many of them are thrown into the eye. Those which enter the eye cause us to see the image in the glass.

Fig. 86.

The Image of a point. — Now, if the rays of light, which form the image in the glass, were visible, the person would be able to trace them back from the eye, converging toward the points on the glass from which they are reflected, and they would appear as if they came from points in the image behind the glass.

This will be better understood by means of Fig. 87. Let M N represent a plane mirror. From the point A, numberless rays fall upon the mirror, some of which, after reflection, will enter the eye supposed to be at O. Two of these rays are represented in the figure. The eye will receive these rays as if they came from the point a, and this point a is the image of the point A, from which the rays proceed.

Images by Plane Mirrors. — We are now prepared to see how looking-glasses make such perfect images of all objects placed in front

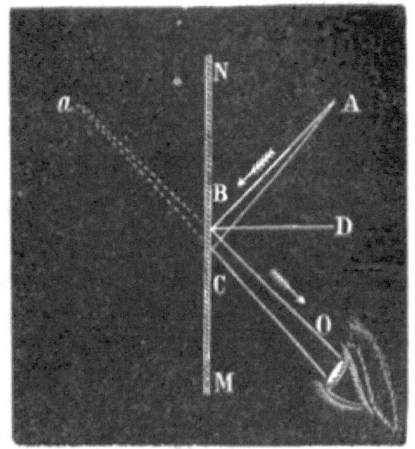

Fig. 87.

of them. Suppose an arrow, A B, placed before a mirror (Fig. 88). Let us construct its image. From the vast number of rays which go from A to the glass, select two which fall upon it very near together, at f and g. By making the angles of reflection equal to the angles of incidence, we find the reflected rays taking the directions f P and g O. Now, if the eye be placed at E, it will receive these reflected rays as if they came from the point a. Again, select two rays, which, going from the other end of the arrow B, strike the mirror at points near together at c and d, so that after reflection they can enter the same eye at E. These rays will appear to have come

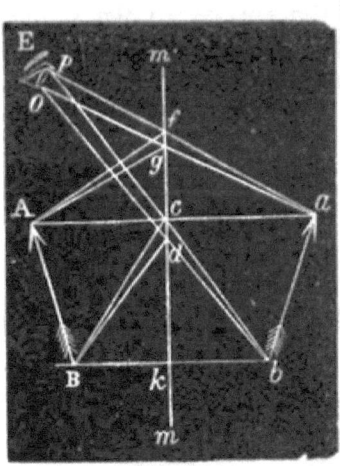

Fig. 88.

from b. From all points between A and B, rays of light will go to the mirror; and, being reflected, will enter the eye at E, and appear to have come from points between a and *b.* The image of the arrow, A B, will thus be seen at *a b*. We may describe this image thus: *The image made by a plane mirror is always behind the mirror, just as far as the object is in front of it, of the same size as the object, and erect.*

72. If an object be placed in front of a concave mirror, the position and size of the image will depend upon the distance of the object from the mirror. We will notice three well-marked cases: —

1st, When the object is beyond the center of curvature.

2d, When the object is between the center of curvature and the principal focus.

3d, When the object is between the principal focus and the mirror. (*G.* 519–521.)

Images are formed. — The brilliant inner surface of a silver spoon shows the image of a person who looks upon it, but it will be curiously different from his image seen in a looking-glass. It is very small; it is inverted; and, moreover, by careful attention, the person sees his picture standing in the air between himself and the surface of the spoon. Nor is this all: the picture in the air will grow larger or smaller, or it may disappear altogether, as the spoon is moved toward or from the face of the observer.

If a spherical concave mirror of small curvature be at hand, a beautiful experiment will illustrate its power to form images. Let a lighted candle be placed in front of a concave mirror. The mirror will receive the light, and reflect the rays upon the wall above the window; and, if its distance from the candle is just right, a fine image, much larger than the candle, and with the flame downward, will be seen upon the wall or screen (Fig. 89).

The Images of Points. — How is this beautiful effect produced? Can we find the images of points of the object by tracing the reflected rays which produce them? Let M N (Fig. 90) represent a section of the concave mirror, and suppose an arrow, A B, in front of it. Select two rays of light which, going from the point A, fall upon the mirror at D and E. After reflection they will cross each other at A'.

Fig. 89.

Again select two rays, which, going from the point B, fall upon the mirror at the points H and F. After reflection they will cross each other at B'. Other points in the object

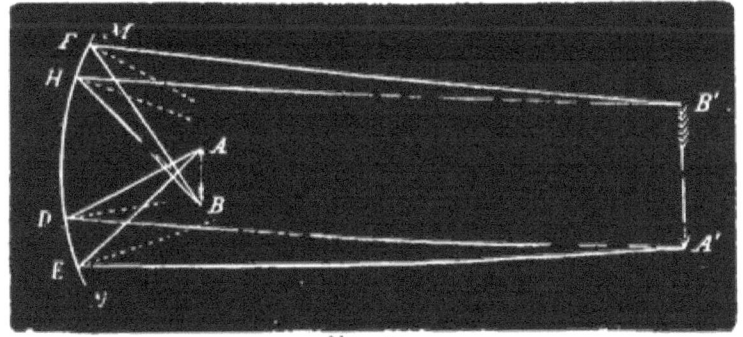

Fig. 90.

will send rays to the mirror, which, after reflection, will cross each other at points between A' and B'. In this way a large and inverted image is made in the air at A' B'.

Case 1: Object beyond the Center. — Let M N (Fig. 91) represent a section of the concave mirror, whose center of curvature is C, and whose principal focus is F. Suppose an arrow, A B, to be put in front of the mirror, beyond the center of curvature. A perfect image of the arrow will thus be formed at A' B'. In this case we observe that *the image is between the center of curvature and the principal focus, inverted, and smaller than the object.*

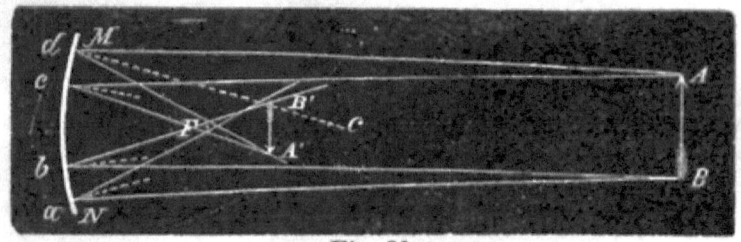

Fig. 91.

This case was illustrated by the experiment with the silver spoon.

Case 2: The Object between the Center and Focus. — Now let us suppose that in this same Fig. 91 an arrow B' A', with its head pointing downward, is placed between the center of curvature and the principal focus. The image of the arrow will be formed at A B. In this case, we observe that *the image will be beyond the center of curvature, inverted, and enlarged.*

This case was illustrated by the experiment with the candle, Fig. 89.

Object nearer to the Focus. — When the object is gradually moved from the center toward the focus, the image will rapidly move farther and farther away, until, when the object has reached the focus, the image will be at an infinite distance *in front of the mirror*, and of course invisible. But let the object be carried a little farther, so as to be between the focus and the mirror, and the image suddenly leaps from its distant place in front of the mirror, to a position *behind* it.

Case 3: The Object between the Focus and the Mirror. — To illustrate the formation of this image behind the mirror, let A B (Fig. 92) represent an object between the focus F, and the mirror M M'. Tracing the rays from A and from B, we find that after reflection they enter the eye at E, and appear as if they came from a and b. In this case we observe that *the image is behind the mirror, erect, and larger than the object.*

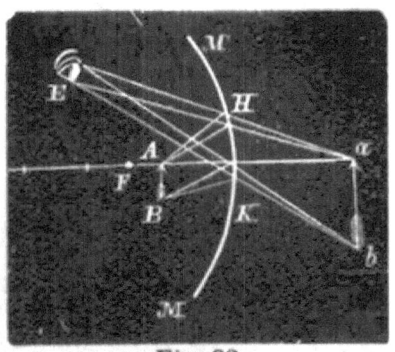

Fig. 92.

73. The images formed by convex mirrors are always behind the mirror, erect, and smaller than the object.

Images by Convex Mirrors. — The convex surface of a silver spoon will serve, in a homely way, to illustrate the effects of a convex mirror. A person looking upon it will see his own image, apparently in the metal of the spoon, erect, but very small. The diagram (Fig. 93) shows how these images are formed.

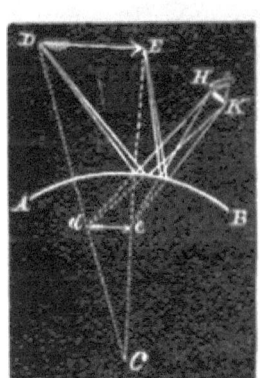

Fig. 93.

74. The law of reflection applies to rough surfaces as well as to mirrors.

Reflection from Rough Surfaces. — Reflection from a rough surface is represented in Fig. 94. The light is scattered in all directions. Yet the law of reflection is not transgressed. Every ray must be thrown in such a direction that the angle of reflection is equal to the angle of incidence. But on a rough surface, like that seen in Fig. 94, the reflecting points are not in regular order, and for this reason the reflected rays are not.

SECTION III.

ON REFRACTION.

75. When light passes from one medium into another of different density, it is refracted, obeying the following laws: —

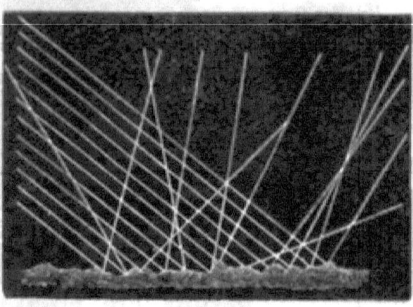

Fig. 94.

1st, In passing into a denser medium, light is bent toward a perpendicular to the surface at the point of incidence.

2d, In passing into a rarer medium, light is bent from the perpendicular. (*G*. 523-526; *A*. 807.)

The Experiment. — Through a small opening in the shutter of a darkened room, let a beam of sunlight enter, and fall obliquely upon the surface of water held in a glass vessel (Fig. 95). If the water has been made turbid by the addition of a little soap, and the air above it misty by sprinkling into it the dust of a chalk-brush, the beam of light will be distinctly seen in both, absolutely straight except at the surface of the water, where it will be very considerably bent.

Fig. 95.

Definition. — This change in the direction of a wave on entering a second medium is called REFRACTION.

The First Law of Refraction. — If now a perpendicular, F E, be erected to the refracting surface at the point of incidence, B (Fig. 96), we see that the rays A B, instead of moving in a straight line onward to C, will be bent toward the perpendicular. Water is denser than air. In going from the *rarer to the denser* medium, the light is bent *toward* the perpendicular.

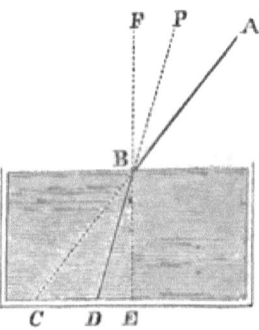

Fig. 96.

The Second Law of Refraction. — Let us suppose that D B represents a beam of light going from the water into the air at B; it will take the direction B A, instead of going on in a straight line toward P, being bent from the perpendicular F E. In passing from the *denser* medium *into the rarer*, the light is bent *from* the perpendicular.

Many phenomena in nature may be explained by reference to these principles. When, for example, an oar is dipped into clear and quiet water, it appears broken at the surface (Fig. 97). The light comes to the eye from all points of the oar. From that part which is above water it comes in straight lines through the air, but from the part under the water the light coming up into the air is bent at the surface. The eye which receives these bent rays traces them back in *straight lines;* and the oar, from which they come, is thus made to appear to be where it really is not

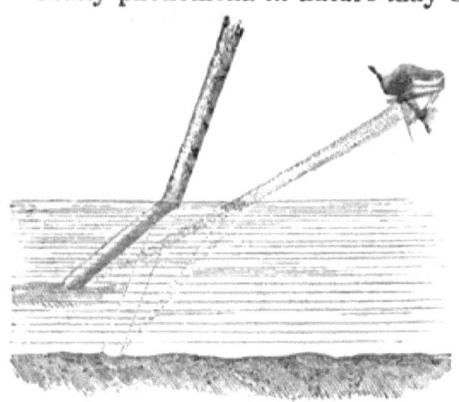

Fig. 97.

76. Some substances refract light more than others. Their relative refracting powers are indicated by certain numbers, which are called INDICES OF REFRACTION.

The Index of Refraction. — We may best explain the meaning of this term by means of the following diagram. Suppose a small beam of light, R I (Fig. 98), to be passing from air into water. It will be bent at I, and go on in the direction of I S. Now, with the point I as a center, and with any convenient radius, describe a circumference. Let a perpendicular, P I, be erected to the surface of the water at the point I, and from the points S and o let lines be drawn perpendicular to P I. The perpendicular from o is the sine of the *angle of incidence*, and that from S is the sine of the *angle of refraction*. If now we measure the lines o t and P S, and divide the length of o t by that of P S, we will obtain a quotient which is called the INDEX OF REFRACTION.

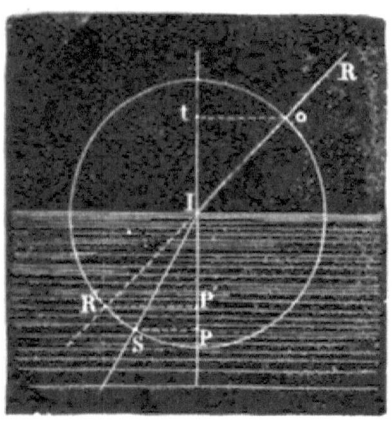

Fig. 98.

The Law. — Now, R I might make a larger angle of incidence or a smaller one, i.e., the light might enter more or less obliquely into the water, but still the *quotient* found by dividing the sine of the angle of incidence by the sine of the angle of refraction would be the same.

$$\frac{\sin. I}{\sin. R} = n.$$

For the same media the index of refraction is a constant quantity.

The Index varies with the Media. — In passing from air into glass, the light is bent more than when passing into water. Hence the index of refraction is larger.

The larger the index of refraction, the greater the refracting power of the substance.

For water, the index of refraction is always 1.336; for crown-glass, the index is 1.58; for the bisulphide of carbon, it is 1.673.

77. The effects of lenses are explained by the principles of refraction.

A convex lens collects the rays of light which pass through it.

A concave lens separates the rays which pass through it. (*G.* 540, 541, 544.)

Lenses. — A lens is a transparent body bounded by surfaces, one at least of which is curved. Six different varieties are used in the arts. They are usually made of glass, and their shapes are represented by sections in Fig. 99

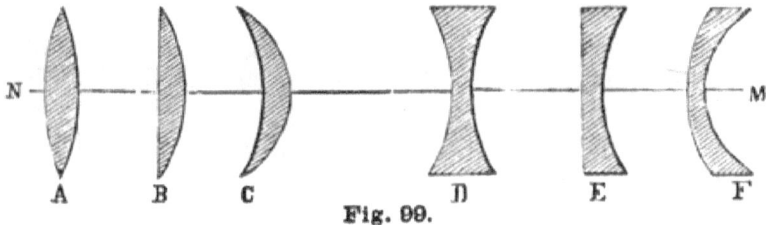

Fig. 99.

The *double convex lens*, A, is bounded by two convex surfaces.

The *plano-convex*, B, is bounded by surfaces, one of which is convex and the other plane.

The *converging concavo-convex*, C, has one surface convex and the other concave, the convexity being greater than the concavity.

The *double concave lens*, D, has two concave surfaces.

The *plano-concave lens*, E, has one surface concave and the other plane.

The *diverging concavo-convex lens*, F, has one surface convex and the other concave, the convex surface being less curved than the concave surface.

The first three of these varieties, A, B, C, are *convex* lenses; the others, D, E, F, are *concave* lenses.

Effect of Convex Lenses.—If a double convex lens is held in the path of a sunbeam, especially in a darkened room, the rays will not come out of it parallel: they will be so bent as to all come to one point,—the principal focus of the lens, as we are shown by Fig. 100.

We may suppose the rays to *go from* the focus to the

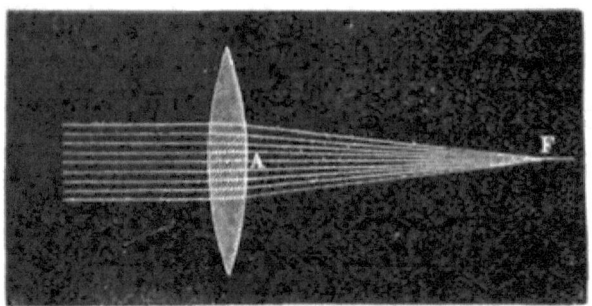

Fig. 100.

lens: these rays (F A, Fig. 100) are diverging, and, as the figure represents them, they will be parallel after going through the lens.

But diverging rays are not always made parallel; indeed, they will not be unless they start from the principal focus. If they start *from a point between the focus and the lens*, they will diverge after going through, but the *divergence will be less* than before. On the other hand, if the rays start from a *point farther than the focus* from the lens, they will be *converging* after refraction. This case is beautifully shown in Fig. 101.

We should find, by experiment, that in all cases the rays after refraction will be nearer to each other than before. *The general effect of the convex lens is to collect rays of light.*

The plano-convex lens and the converging meniscus (C, Fig. 99) will have the same effect, but in a less degree.

Foci.—The point F in Fig. 100 is the *principal focus* of the lens: it is the focus of rays which are *parallel* to the

NATURAL PHILOSOPHY. 189

axis. The distance of this point from the lens will depend upon the curvature of the lens, and upon the index of refraction. If the two surfaces of the lens are equally curved, and it be made of glass whose index of refrac-

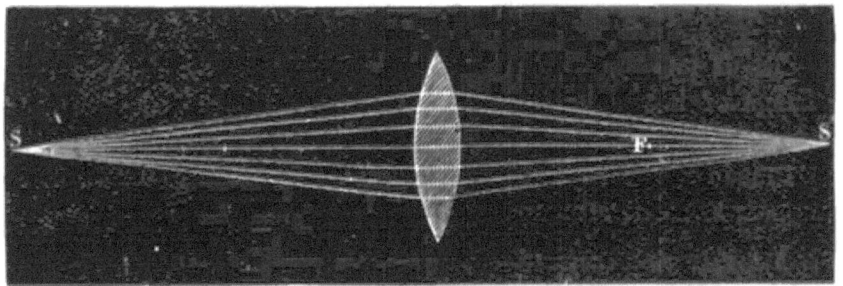

Fig. 101.

tion is 1.5, then the principal focus will be at the center of curvature.

The points S and S' in Fig. 101 are *conjugate foci*. They are so related that the light radiating from either one will be collected into the other.

Effect of Concave Lenses. — The rays after passing through a concave lens are separated instead of being collected.

This effect is **well shown in Fig. 102**, which represents a

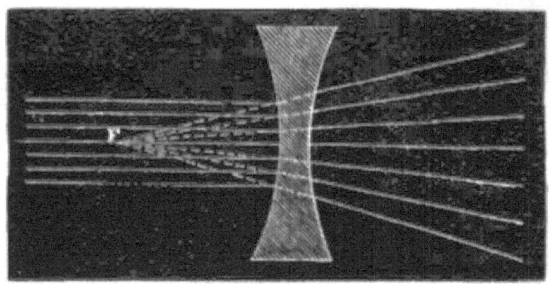

Fig. 102.

double-concave lens refracting parallel rays of light. They are supposed to enter the lens on the side F, parallel to each other, but on coming out on the other side they are diverging. Diverging rays are made more diverging, while converging rays are made less converging. In all cases, *rays refracted by a double-concave lens are separated.*

The plano-concave lens and the diverging concavo-convex lens have the same effect, but in a less degree.

78. If an object be placed in front of a convex lens, an image of it will be formed on the other side of the lens. To explain this, remember that the image of any point will be made where rays of light going from it either meet, or appear to meet, after refraction.

Images are formed.—If a convex spectacle-glass is held in front of a window, at considerable distance, and a sheet of white paper is put in front of it, the light from the window will go through the glass, and fall upon the paper. If the distance from the glass to the paper be just right, a very small but very perfect image or picture of the window will be seen upon it.

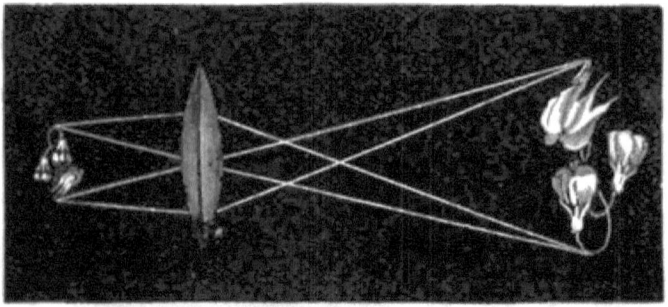

Fig. 103.

If a good double convex lens, three or four inches in diameter, be at hand, a very beautiful experiment may be made. Through an opening in a shutter of a darkened room, admit a beam of sunlight. Into this beam put any small, transparent object, it may be a picture painted on glass, or, quite as well, a wing of the dragon-fly, or a delicate flower. If now the lens be moved back and forth in front of this object, until just the right distance is found, a very large and perfect image will be seen inverted upon the opposite wall of the room. (Fig. 103.)

Explanation. — Now let us see how these beautiful effects are produced.

Suppose an arrow N S, (Fig. 104), placed at some distance in front of a convex lens, M, whose centers of curvature are f and f'. Two rays of light from the point N, passing through the lens, will be refracted so as to cross each other at the point n. This point, where rays of light meet *after refraction, is* the image of the point N, *from which they came.* The rays from the point S of the object, after refraction, cross each other at s, and form an image there. From points between N and S, rays of light going through the lens will be collected on corresponding points between n and s, and thus a perfect image will be made inverted at n s.

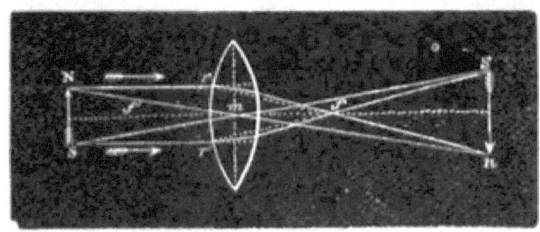

Fig. 104.

In this way it is easy, by a diagram, to illustrate the formation of all images by lenses.

79. The place and size of the image will depend on the distance of the object from the lens. Four typical cases may be studied : —

1st, The object is placed at twice the focal distance from the lens.

2d, The object is more than twice the focal distance from the lens.

3d, The object is beyond the focus, but less than twice the focal distance.

4th, The object is at less than the focal distance.

Case 1: The Object twice the Focal Distance. — Suppose the lens to be one whose focus is at the center of curvature, and that the object is just twice that distance from the lens, as shown by the arrow N S (Fig. 105). Two rays of light from the top of the arrow go through the lens, bending according to the laws of refraction, and cross each

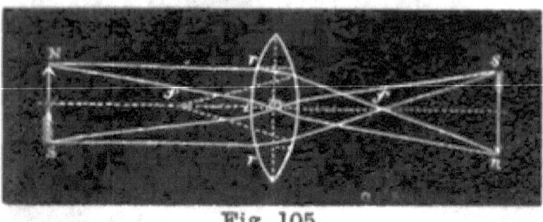

Fig. 105.

other at the point n. Two rays from the bottom of the arrow go through the lens, and cross each other at the point s. Join the points n and s, and $n\ s$ represents the image that is formed. This *image will be at twice the focal distance on the other side of the lens, of the same size as the object, and inverted.*

Case 2: The Object farther away. — This case is represented by Figs. 103 and 104. Suppose that in front of the lens M, an arrow, with its head downward, represented by $n\ s$, is placed at more than twice the focal distance from the lens. Two rays from the arrow-head, after refraction, will be found to cross each other at N; two rays from s will, after refraction, cross each other at S. *The image N S is on the other side of the lens, at a less distance, smaller than the object, and inverted.*

Case 3: The Object at a less Distance. — If, in Fig. 104, we suppose N S to represent the object, outside the focus, but at less than twice the focal distance, its image will be found at $n\ s$. In this case *the image will be at a greater distance on the other side of the lens, larger than the object, and inverted.*

Case 4: The Object between the Focus and the Lens. — One more case remains to be considered. Suppose the object to be between the focus and the lens. Let M N (Fig. 106) represent a lens whose focus is at C, and let the object A B be placed between this point and the lens. An attentive examination of the figure shows that the rays of light from the point A are diverging after refraction. And, since they can never meet, it is clear that no image can be formed on that side of the lens; but if an eye at E receive these rays they will produce the same effect as if they came from A'. In like manner, the rays from B, entering the eye at E, will seem to have come from B'. Hence an image will seem to be formed at A' B'. *This image will be on the same side of the lens as the object, erect, and larger than the object.*

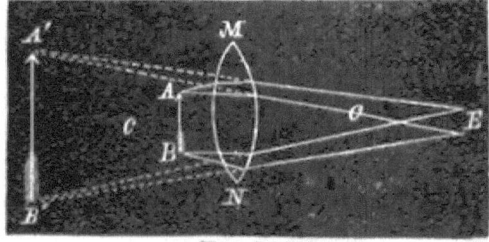

Fig. 106.

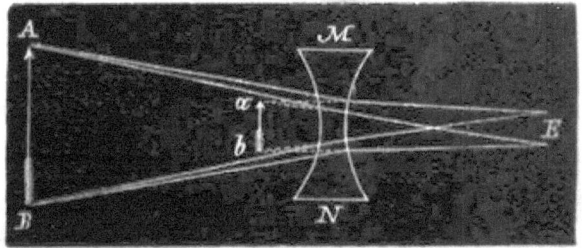

Fig. 107.

80. Images are also formed by concave lenses. They are on the same side as the object, smaller, and erect.

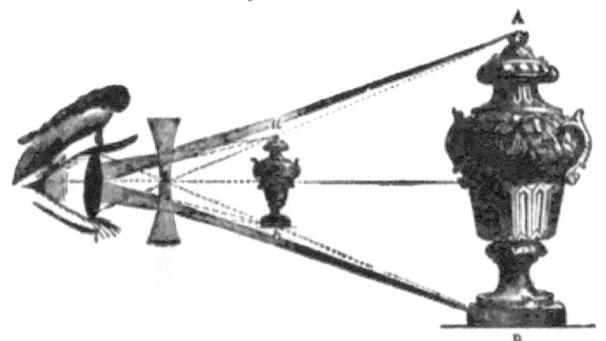

Fig. 108.

Image by Concave Lens. — Suppose an object, A B (Figs. 107 and 108), in front of a concave lens M N. Rays of light from A, after refraction, diverge as if they had come from *a;* rays from B, after refraction, diverge as if they had come from *b;* the image will thus appear to be made at *a b.* This *image is on the same side of the lens, smaller than the object, and erect.*

SECTION IV.

ON DISPERSION.

81. Prisms refract light; they also disperse it. They separate white light into rays of seven different colors, — viz., violet, indigo, blue, green, yellow, orange, and red, — together with invisible rays of heat and actinism. (*G.* 522, 523; *A.* 812, 896.)

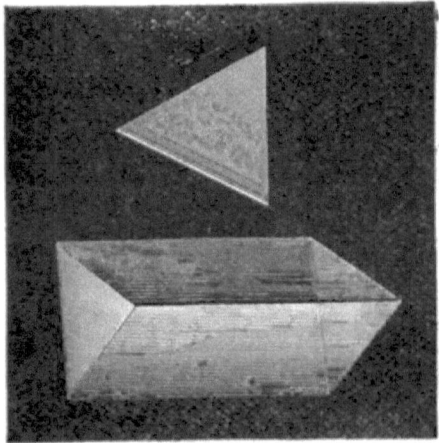

Fig. 109.

Prisms. — Any transparent body, two of whose sides are inclined toward each other, is a PRISM. The most common form of the prism is a triangular piece of glass (Fig. 109). A water-prism may be made by taking a three-cornered vessel, with glass sides, and filling it with water. Other fluids may be used in place of water.

Prisms refract Light. — Light, in passing through prisms, must obey the laws of refraction. In Fig. 110 the triangle *m n o* represents a section of a prism. A ray of light striking its surface at *a* will be bent toward a perpendicular on entering, and from a perpendicular on emerging, finally taking the direction *b c.* To the eye at *c,* this light would seem to come from the object at *r* instead of L.

NATURAL PHILOSOPHY. 195

Prisms decompose Light. — The white light that comes from the sun, or from other luminous bodies, is really made up of seven different kinds of light. The way in which

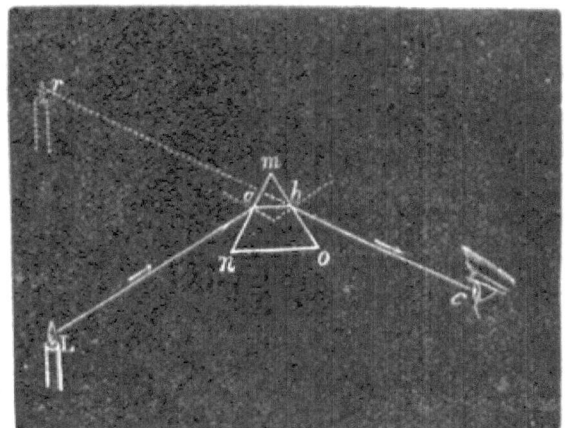

Fig. 110.

Sir Isaac Newton made this great discovery is shown in Fig. 111. In the window-shutter S, of a darkened room,

Fig. 111.

he made a small hole, and placed behind it a prism, P, so that the beam of sunlight could fall obliquely upon one of its sides. Without the prism the beam of light would have gone straight forward to the floor, where it would have made a round white spot; but, being refracted by the prism, it was thrown upon the screen E, and an oblong image containing *seven different colors* appeared. These colors were, in order from the top of the image, violet, indigo, blue, green, yellow, orange, and red.

These colors are separated, because the prism has power to bend some of them more than others. The violet rays are bent most, the red rays least.

The oblong image upon the screen is called the SOLAR SPECTRUM. This separation of the white light into its constituent rays of different refrangibility is called DISPERSION. The power of a prism to separate the color of white light is called DISPERSIVE POWER.

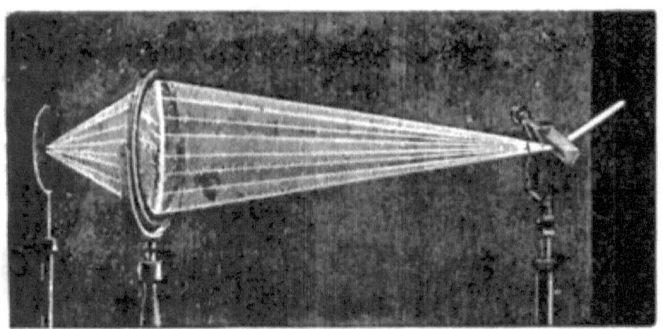

Fig. 112.

Recomposition. — The prism in this way enables us to analyze white light, or to find out the colors of which it is made; and now, if by any means we can unite these seven colors, we shall produce white light again. This can be done by using *any instrument which collects rays of light*. If the rays fall upon a concave mirror, they will be reflected to a focus, which will be a *white* spot. If the rays are received upon a double-convex lens, they will be refracted to a focus (Fig. 112), and this focus also will be *white*. Sir Isaac Newton collected the rays by using a second prism, exactly like the first, but placed beside it so as to bend the rays in the opposite direction: the image on the screen was *white*.

82. A pure spectrum formed by sunlight or by starlight is crossed by a great many fine black lines, while the spectra formed by light from artificial sources are crossed by different-colored bright lines. (*G.* 562, 563.)

A Pure Spectrum. — In the spectrum obtained from a beam of considerable width, as that which comes through a

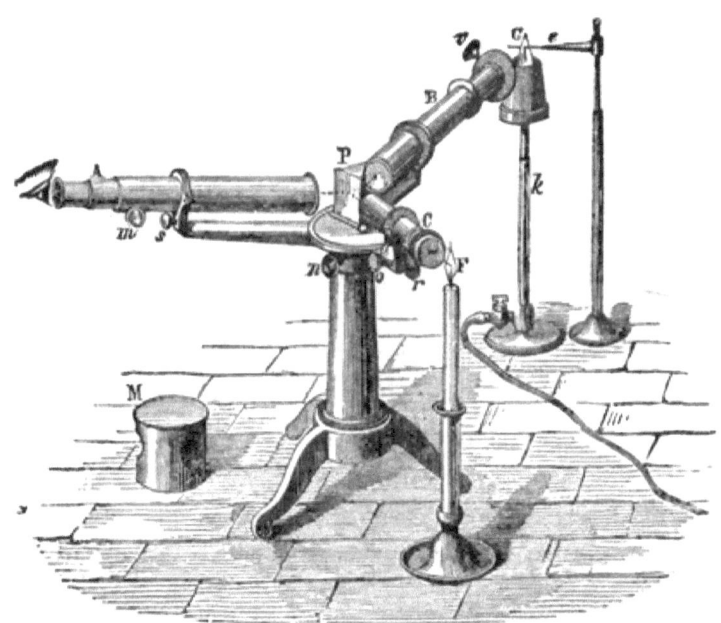

Fig. 113.

round hole in the shutter, the colors overlap one another, and, on this account, are more or less mixed. In order to get the colors pure, the opening must be in the form of a *narrow slit*.

The Spectroscope. — The spectroscope is an instrument used in the study of spectra. It is shown in Figs. 113 and 114.

The rays of light from any source enter the tube B through a narrow slit in the end, and are made parallel by a convex lens at the other end. They then pass through the prism P, are decomposed, and the colors pass through the telescope A to the eye of the observer.

The black Lines. — The whole length of the solar spectrum, when seen by the naked eye, seems to be colored: it is a *continuous spectrum;* but, when seen through a spectroscope, a great many fine *black* lines are found to cross it, as if a delicate brush, dipped in the purest black, had been drawn across it by a skillful artist. It is a *line spectrum.* A

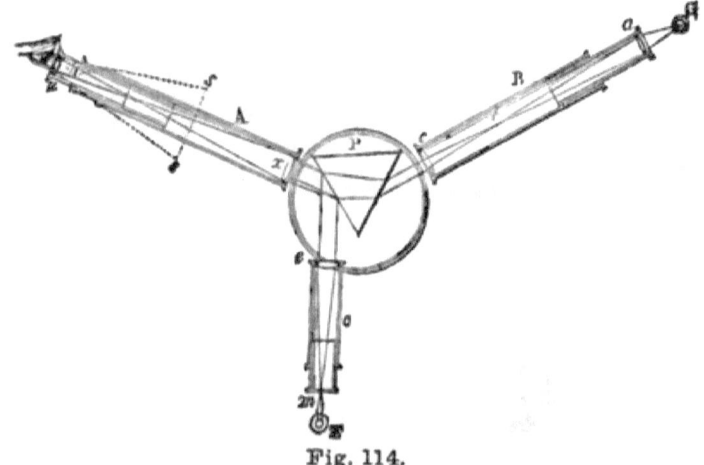

Fig. 114.

beam of sunlight always gives the same set of lines, holding the same relative position in the spectrum. A beam of starlight gives a different set, and the light from different stars gives each a set of its own. These lines are usually called FRAUNHOFER'S LINES, in honor of him who first examined them carefully.

The bright Lines. — When the light from an artificial source is passed through a prism, and its spectrum is seen in a spectroscope, no black lines are visible, but instead of these there will be seen lines of exceeding brightness, and of different colors. The colors of these lines, and their places

NATURAL PHILOSOPHY. 199

in the spectrum, will depend upon the substance whose flame gives the light. If, for example, a little common salt be burned in a hot gas-flame, a yellow line of surprising brightness will *always* appear in the yellow part of the spectrum, while the metal potassium in the flame will *always* give two lines, one of a brilliant crimson color in the red end of the spectrum, the other a beautiful blue line away off in the violet end. Each substance gives a set peculiar to itself.

Four of these spectra are represented in Fig. 115. No attempt is made to show the colors of the lines, but their relative positions are seen marked by a scale along the top.

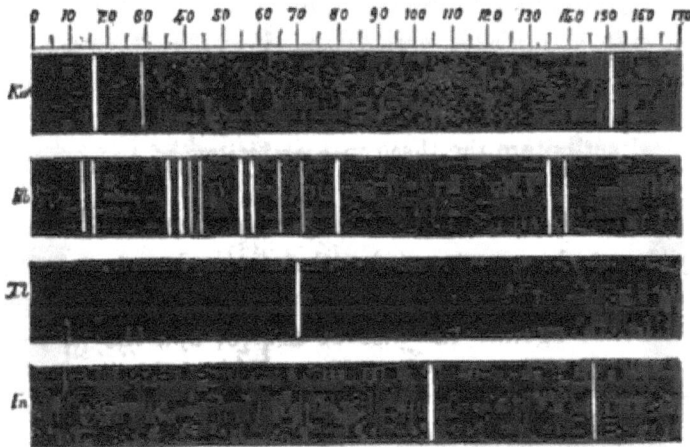

Fig. 115.

The scale is pictured in the spectroscope along with the spectra, by means of the tube C, and marks the places of the lines with accuracy. The upper spectrum in the cut is that of potassium. Those of rubidium, thallium, and indium follow.

The art of detecting the presence of substances by means of the spectra obtained by igniting them is called SPECTRUM ANALYSIS.

Invisible Parts of the Spectrum. — A sensitive thermometer, held in the colors of the spectrum, shows the presence of heat; and a curious fact was first discovered by

Herschel, who found that the heat was more and more intense toward the red end, and, further, that the thermometer revealed the presence of heat at some distance below the red where *there was no color* at all.

There is an invisible part of the spectrum, less refracted than the red, which shows itself as heat.

Another Experiment was made by Ritter, who let the spectrum fall upon a sheet of paper covered with solution of silver-nitrate. The pure white paper *was blackened* by the spectrum. And, what was more surprising, Ritter found that this effect was produced upon the paper to quite a little distance beyond the violet where *there was no color.*

There is an invisible part of the spectrum, more refracted than the violet, which shows itself by chemical action.

The Energy of the Sunbeam. — Heat, colors, and chemical action are the three manifestations of the energy in the sunbeam. The sunbeams themselves are only undulations from the sun. But these undulations are compound. They are made up of a multitude of simple waves differing greatly in their lengths and periods. These compound waves break against the face of nature, and the molecules of bodies are put in motion by them.

The energy of this molecular vibration *at certain rates* is *heat*, at certain more rapid rates is *color*, and at still more rapid rates is *actinism*.

83. Heat tends to diffuse itself equally among all bodies.

The amount of heat which a body can radiate depends upon its temperature, its nature, and the condition of its surface.

The equal Diffusion of Heat. — If two bodies, one cold, the other hot, be placed near each other, it will in a short time be found that both are equally warm. The cold body has received more heat, the hot body has parted with some that it had. What is thus true of *two* bodies is true of *all*. Bodies are constantly giving and receiving heat. Those which part with more than they receive from others grow colder; those which receive more than they give grow warmer. Ice, for example, is giving heat to all bodies around it; it is at the same time receiving heat from them in return. Ice will actually warm a body which is colder than itself, because it will give more heat than it gets in return; it will be melted by a body warmer than itself, because it receives more than it gives.

Transmission of Heat-Rays. — Just as light passes more freely through some bodies than through others, so the dark rays of radiant heat pass through different bodies with different degrees of facility. Those bodies through which heat passes most freely are said to be *diathermic*, while those through which it can go with the greatest difficulty are said to be *athermic*.

Heat from different sources is transmitted in different degrees through the same substance. It is, for example, a familiar fact, that the glass of our windows allows the heat of the sun to enter our rooms, while it prevents the heat of the stove from going out.

Rock-salt is the most diathermic substance known; it allows heat from all sources to pass through .t with the greatest freedom.

Depends on Temperature, Nature, Condition. — The higher the temperature of a body, the more heat it can radiate. Moreover, bodies of different substance, when at the same temperature, give off different quantities of heat in the same time. Thus iron is a better radiator than gold. Still further, the same body, at the same temperature, with a *rough* surface will radiate much faster than when its surface is smooth. The rough surface of a cast-iron stove, for example, is a better radiator than if it were polished.

84. Drops of rain may decompose the sunlight: in this way the rainbow is produced. The primary bow consists of bands of the seven colors of the spectrum, arranged in parallel arches, with the red band on the outside.

In the secondary bow the order of the colored arches is changed, the violet being on the outside.

The Primary Rainbow. — This most beautiful phenomenon is produced by the action of rain-drops; they decompose the sunlight, and send its rich colors to the eye.

To understand this action, suppose the circle, whose center is at C (Fig. 116), to represent a section of a drop of water.

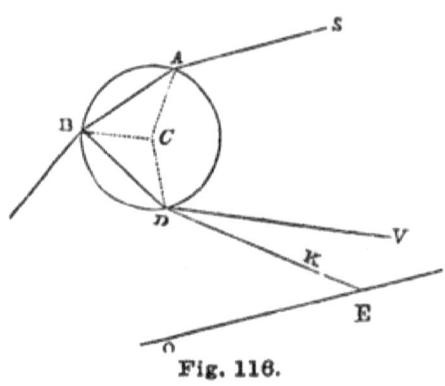

Fig. 116.

Rays of sunlight (S A) falling upon the upper part of the drop will be refracted to the point B. At this point a part of the light will pass out into the air again, but another part will be reflected by the inner surface of the water, and strike the surface at another point, D. The light which here goes out of the drop into the air will be again refracted. The light will not only be refracted, in its passage through the drop: it will be, at the same time, *decomposed*. On coming out of the water the red ray, bent least, will take a direction represented by D E; the violet ray, bent most, may be represented by D V; and all the other colors of the spectrum will be found between these.

The red Band is on the Outside. — Now, it is quite clear that if the person were standing upon the ground in the direction of D E, so that the red rays from this drop would enter his eye, the violet rays, and indeed all the other colors, would go over his head. To him this drop of water would appear red. Another drop, some distance *below* this one, would send violet rays into the same eye. Between the drop which sends the red, and that which sends the violet, there would be others from which the eye would receive the other colors of the spectrum. (Fig. 117.)

NATURAL PHILOSOPHY. 203

Hence, when a shower of rain is falling, and the sun is at the same time shining in the opposite part of the sky, so that a person looking toward the shower will have his back turned toward the sun, he will see the seven colors of the spectrum painted upon the cloud in order, with red at the top and violet at the bottom.

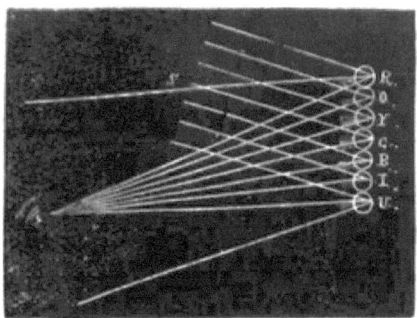

Fig. 117.

The Colors are in the Form of an Arch. — Now, suppose a line drawn from the sun through the eye of the observer, and straight onward until it reaches a point O (Fig. 116), directly under the drop C, which sends the light to the eye. If this drop sends a red ray to the eye, then all others, which like this are opposite the sun, and whose distances from O are the same, will also give red rays. If the arc of a circumference be drawn with O as a center, and with a radius C O, all drops along this circumference will be equally distant from the center O, and will therefore give red rays. The red part of the rainbow is, for this reason, a circular arch, and, for the same reason, the other colors are parallel arches below the red.

The Secondary Bow. — Outside of the bow just explained, another, the secondary bow, is often seen. Its colors are more dim, and their order is reversed, the violet being at the top and the red at the bottom.

To explain the primary bow we trace rays of light falling upon the top of the drops of water. But drops of rain in the air are entirely covered with light, and, to explain the secondary bow, we may trace the rays which fall upon their

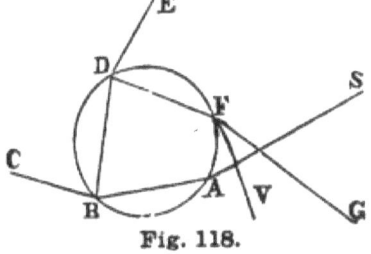

Fig. 118.

lower parts. The diagram (Fig. 118) illustrates this. A beam of light, S A, is refracted at A, reflected at B, and again at D, and refracted at F, finally entering the eye in the direction of G. If F G represents the red ray, then F V may represent the violet ray which is more refracted. Hence the drop which sends the violet ray to the eye must be above that which sends the red; and the violet band is on the outside.

85. Bodies are of different colors, only because they decompose the sunlight, and reflect different parts of it to the eye. The various colors of the sky and the clouds are due to the decomposition of the light which comes through them from the sun. (*G*. 555; *A*. 913.)

The Color of Bodies. — The sun sheds a flood of pure white light upon all bodies alike. This white light is decomposed at their surfaces. Some of its colors are transmitted or absorbed by the body, while the others are reflected to the eye. One body is red because it decomposes the sunlight, and reflects the red rays; another is blue, because it reflects only blue rays. The foliage of trees in the springtime receives the sun's white light, decomposes it, and reflects only the green rays. The petals of the violet decompose the sunlight to share with us the beautiful colors of the spectrum; it reflects the colors of the violet end, and keeps to itself those of the other. A body which reflects all the color of the light it receives is white; one which reflects none is black.

That which is absorbed becomes heat or chemical action.

The Color of the Sky. — The sky, when free from clouds, is blue, because the particles of the atmosphere reflect blue rays of light. If the thin air could not reflect light at all, the sky would appear black: if it reflected it without decomposition, it would be white. The white sunlight falls upon its molecules, is decomposed by them, and only those rays which make up the delicate blue color of the sky are reflected to our eyes.

The Color of the Clouds. — The clouds both reflect and refract the sunlight, and all their varied colors are due to the decomposition thus produced. There can be no more gorgeous display of colors than we often see upon the clouds of the morning and the evening sky. What grand and diversified effects to be produced by means of such simple materials as light, water, and air!

86. The colors of the spectrum may be also produced by the interference of light-waves.

This furnishes a means of measuring the wave-lengths, which are found to vary from .0000266 of an inch for red to .0000167 for violet.

"Diffraction fringes" are the effect of interference. (*G.* 627, 630, 631.)

Sound and Light alike. — We have learned that light is the result of vibrations in a very elastic medium called ether. We ought, therefore, to expect that the phenomena of light would, in many respects, be like those of sound and heat. We have found this to be true, their laws of reflection, refraction, and transmission being alike. They are alike also in regard to interference.

Interference of Sound. — If two sound-waves in air cross each other, so as to bring their condensed parts or phases together, they cause a wave whose amplitude is equal to the sum of theirs, and produce a sound of greater intensity. If they come together in such way that the condensed phase of one strikes the rarefied phase of the other, they cause a wave whose amplitude is equal to the difference of theirs, and produce a sound of less intensity.

Thus two sounds may together produce silence.

Interference of Light. — So, too, if waves of light, in the ether, cross each other so as to bring their like phases together, they cause a wave whose amplitude is equal to the sum of theirs, and produce a light of greater brightness; but, if their opposite phases are thrown together, they form a single wave whose amplitude is equal to their difference, and cause a light of less intensity.

Thus two rays of light may together produce darkness.

The Conditions. — Now, examine the conditions of interference more carefully. A wave consists of two phases, and the sum of their lengths is the length of the wave. Of course, then, each phase is just one whole wave-length ahead of the next one behind it of the same name, and just one-half a wave-length ahead of the next one behind it of a different name. If, then, two sets of waves are to interfere with like phases together, their starting-points must be *one wave-length*, or some *whole number* of wave-lengths, apart. To bring different phases together, the distance between their starting-points must be *one-half a wave-length*, or some *odd number* of half wave-lengths apart.

If the two sets have equal *amplitude*, their interference in like phases will give a *double brightness*, but in unlike phases will result in *darkness*.

The Length of Light-Waves. — Light of all colors travels with the same velocity; and, since violet is produced by the most rapid vibrations, the *length* of its waves must be less than for any other.[1] Suppose, now, that two sets of waves start from surfaces very near to each other, but *not parallel:* at some points the distance between the starting-points of two waves will correspond to the wave-length for violet; at others, to the wave-lengths of other colors. The result of the interference of the two sets of waves will be to form, at different points, all the tints of the spectrum. The rainbow colors of the soap-bubble, which so delighted us in childhood, illustrate this most beautifully. The light is reflected from both the outside and the inside surfaces of the thin film; these surfaces are not parallel; and the interference of the two sets of waves gives rise to the colors.

The Measurement. — Now, could we but measure the

---

[1] Wave-length $= \frac{\text{Velocity}}{\text{Rate}}$; or, $L = \frac{V}{R}$. This is true for waves of every name, in water, air, or ether.

thickness of the film at the point where red is seen, we could find the length of the wave for red; and, if at points where other colors appear, we could find the wave-lengths which produce them. Newton actually calculated these minute spaces, although, of course, so frail a thing as a soap-bubble could not be used for the purpose. His plan may be understood from Fig. 119. A very thin layer of air is included between two very smooth glass surfaces, one curved, the other plane. When the glasses are pressed together, a series of rainbow-colored rings, *Newton's rings*, are seen, with a black center at the point a, where the glasses are in contact. If red light alone is used, a series of red rings will be separated by dark spaces.

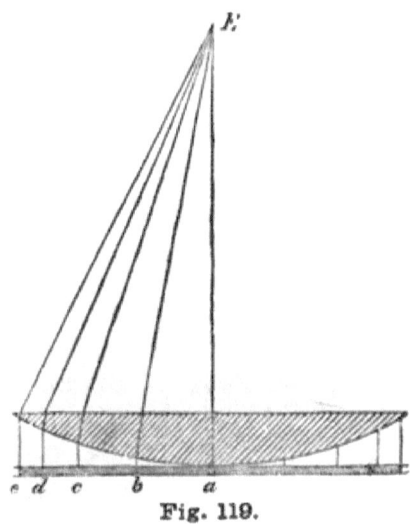

Fig. 119.

Now, these rings are caused by the interference of two sets of waves, one reflected from the lower side of the curved glass, the other from the upper side of the plane glass, meeting at the eye, E. Newton calculated the thickness of the layer of air at points, b c d e, where the rings were seen, and from these thicknesses calculated the length of the waves. The more refrangible colors are produced by shorter waves. The lengths of luminous waves vary between .0000167 of an inch for violet, and .0000266 of an inch for red.

Color depends upon Rapidity of Vibration. — As the pitch of sounds depends upon the rapidity of the undulations of air, so the color of light varies with the rapidity of undulations of ether. A red light is made by the slowest, a violet light by the swiftest, vibrations.

The eye is affected by undulations between the limits of about 392 millions of millions a second for red, to about 754 millions of millions for violet.[1]

Diffraction. — Diffraction is the change which light undergoes when it passes the edge of an opaque obstacle.

Experiment. — A beam of sunlight L (Fig. 120), coming through a narrow slit in the shutter of a darkened room,

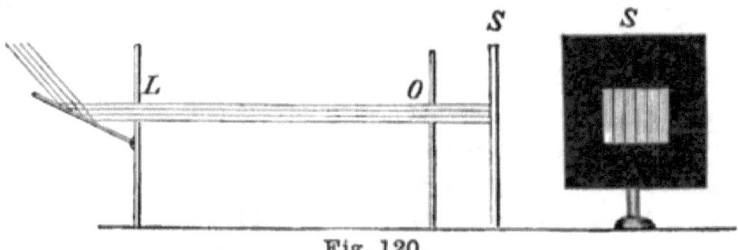

Fig. 120.

is made to pass through a second slit O, at a distance of ten or fifteen feet. When a white screen, S (a front view shown at S), is placed behind this slit, a distance of about four feet, a multitude of colored bands, alternating with dark spaces, will appear upon it.

Explanation. — Diffraction fringes are caused by the interference of light. When one set of waves passes the edge of an obstacle, it starts another set in the ether beyond. Two new sets of waves are in this way started from the opposite edges of the slit in the shutter. These two sets going almost, but not exactly, in the same direction, interfere and give rise to the many colored bands.

Illustrations of Diffraction. — Place two knife-blades edge to edge, and look through the narrow slit between them at the clear, bright sky. Instead of a well-defined, clear, bright space, a great number of very delicate parallel black lines will be seen. The edge of a single blade, or of any thin body, will appear fringed with dark lines, and, under some circumstances, with colored bands of great beauty.

---

[1] Rate for violet $= \dfrac{\text{velocity of light}}{\text{wave-length}} = \dfrac{186,000 \text{ miles}}{.000167 \text{ inch}}$.

One who has been taught to recognize it, will be surprised to find how numerous and common are the various forms of this delicate phenomenon. A lady, on suddenly lifting her eyes to the bright sky, often sees it, through the meshes of her veil, covered with a net-work of rainbows. Who has not wondered at the brilliant colors of the sky, seen through the fine fibres of a bird's feather?

SECTION V.

ON OPTICAL INSTRUMENTS.

87. The microscope, the telescope, and many other instruments, help the eye to see small or distant objects, by forming large and perfect images of them near by, for it to examine.

The eye itself is an optical instrument of the most perfect construction. (*G.* 565, 566, 567, 571, 576.)

The Simple Microscope. — The simple microscope consists of a single convex lens. The lens is held in the hand at a little less than its focal distance from the object. The

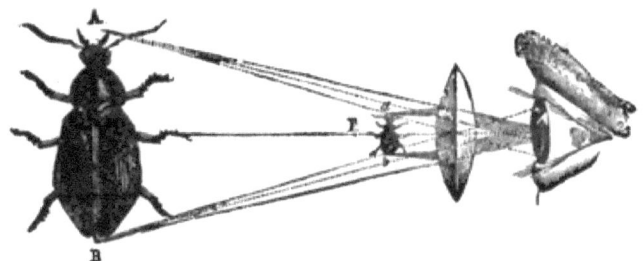

Fig. 121.

eye receives the light which comes from the object through the glass, and sees a magnified image on the other side. In Fig. 121 the object is a small insect, *a b*.

The Compound Microscope. — The operation of the *compound microscope* may be understood by means of a diagram (Fig. 122). It consists of at least two convex lenses.

The lens A, called the *object-glass*, refracts the light from the object O, placed a little beyond its focus, and forms an image inverted at O'. The light from this image is refracted by another lens B, called the *eye-glass;* and, if the rays are received into the eye, they will appear to have come from C D, which is the magnified image of the object.

In a good microscope these glasses are not single convex lenses : each is a combination of two or more. A single convex lens can not give an image sharply defined and free from unnatural colors, but combinations can be made to do this.

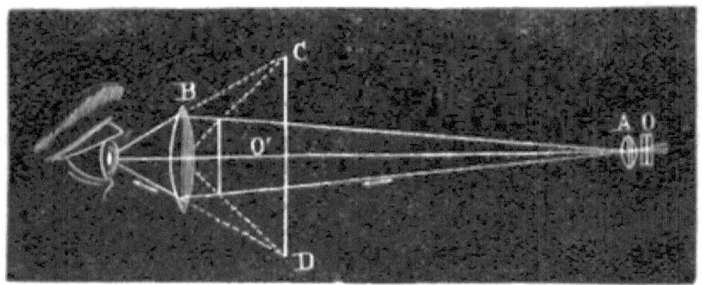

Fig. 122.

By means of this instrument, things otherwise too small to be seen are made visible, and a world of wonderful creations is thus revealed for the study and admiration of man. A drop of water from a stagnant pool is found, by means of the microscope, to be swarming with living creatures, whose forms are perfect and whose appetites are not unlike those of larger animals.

The Telescope. — A telescope is used for viewing distant objects. Sometimes a lens is employed to form an image; sometimes the image is formed by a mirror. In the first case, the instrument is called a REFRACTING TELESCOPE; in the second, it is called a REFLECTING TELESCOPE.

Of the refracting telescope there are three important forms : *Galileo's*, the *astronomical*, and the *terrestrial*.

NATURAL PHILOSOPHY. 211

In Galileo's Telescope there is a double convex object-glass M N (Fig. 123), and a double concave eye-glass, E F.

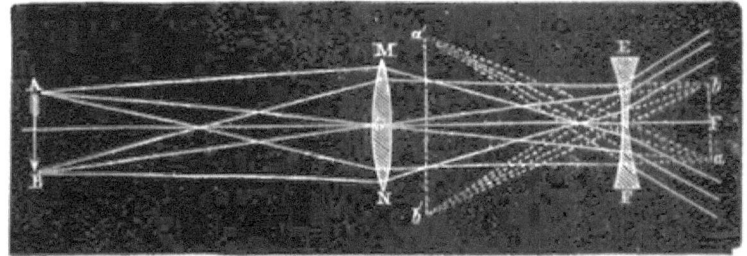

Fig. 123.

Rays of light from the point A of a distant object A B, after passing through the two glasses, diverge as if they came from the point a', while rays from the point B of the object after refraction diverge as if from the point b'. An erect image, $a\ b$, will be seen by holding the eye in front of the eye-glass E F.

The *opera-glass* consists of two small Galilean telescopes placed side by side.

In the Astronomical Telescope, two double convex lenses are used. The object-glass O (Fig. 124) forms a small image $a\ b$ of a distant object A B. The eye-glass,

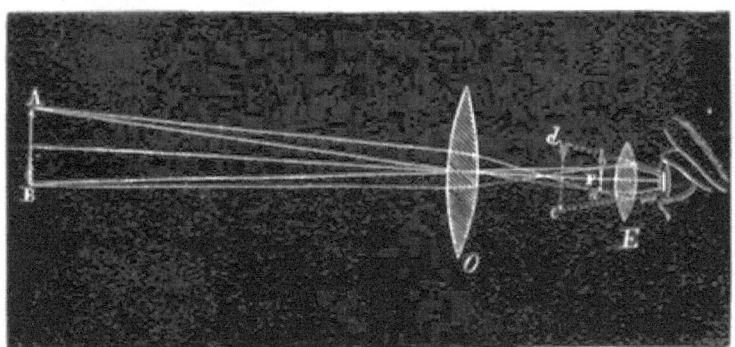

Fig. 124.

E, being so placed that its focus, F, is a little beyond this image, refracts the light, so that it will appear to have come from a magnified image $c\ d$. The course of the rays may be traced in the figure. In this instrument the image is always inverted.

The Terrestrial Telescope is used for viewing distant objects upon the earth. To see them upside down, as in the astronomical telescope, is not desirable: that they may be seen right side up, two convex lenses are placed between the object-glass and the eye-glass. The arrangement of the glasses, and the course of the rays, are shown in Fig. 125. The object-glass O forms a small inverted image, I, of a dis-

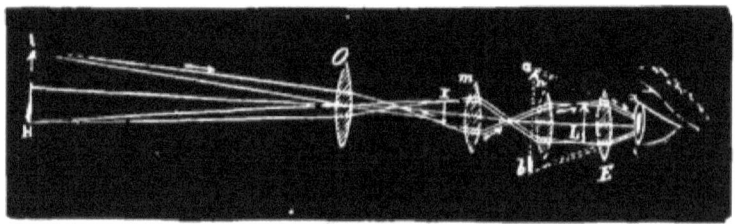

Fig. 125.

tant object, A B, near its focus. From this image the light goes through the two lenses, *m* and *n*, to form a second image L. This image is *erect* with respect to the object, and it is magnified by the eye-glass E, in the usual manner.

Reflectors. — Of the reflecting telescope there are several varieties. In all of them the image of a distant object is

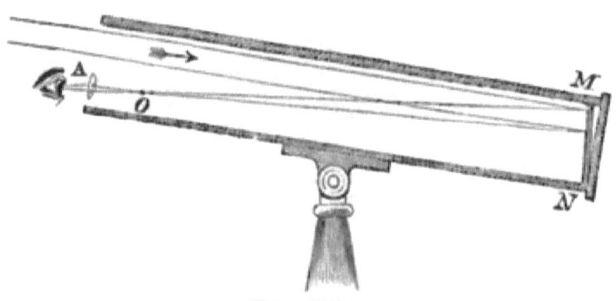

Fig. 126.

formed by a concave mirror, and this image is magnified by a convex eye-glass.

In the **Herschelian Telescope** (Fig. 126), the mirror M N is inclined to the axis of the tube in which it is placed, so that rays of light from a distant object will be reflected to a focus near to one side of the tube at the other end. The observer, looking down into the tube, holds an eye-piece, A, in his hand, through which he views a magnified image.

The Magic-Lantern. — The magic-lantern is an instrument by which the image of a small object, greatly magnified, may be thrown upon a screen.

It consists of a convex lens, with objects highly illuminated by lamp-light placed so near it, that their images are formed far away. Fig. 127 shows a section of the instru-

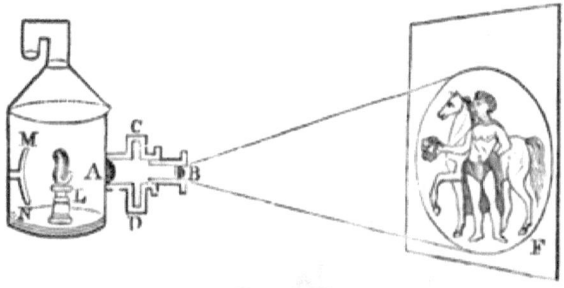

Fig. 127.

ment. Inside of a dark box, a strong light, L, is placed. Behind this light is a concave mirror, M, and in front of it a convex lens A. This lens is at the entrance of a tube which projects from the side of the box. Inside this tube slides a smaller one, in which is fixed another powerful lens. The picture is placed in a slit, C, provided for it in the larger tube, just in front of the first lens. The lamp fills the box with a strong light. The lens A, receiving light directly from the lamp, and reflected from the mirror, condenses it upon the object, and highly illuminates it. The light from this bright object goes on through the second lens to the distant screen, and there forms a large and perfect image.

The stereopticon, so largely used by lecturers and teachers, is a magic-lantern with lenses of superior quality, and in which the lime-light or the electric light is used.

214 NATURAL PHILOSOPHY.

The Camera-Obscura. — The camera-obscura is an instrument by which to form miniature images of objects. It consists of a dark box, a section of which is represented by A B (Fig. 128), containing a screen, S, and having a double convex lens, L, filling an opening in one end. The distance of the lens from the screen may be varied by sliding the tube which carries it back and forth in the larger tube C. The light from the object O is refracted by the lens, and a beautiful image is formed upon the screen. This image is always inverted, and smaller than the object.

The camera may be illustrated by a very simple experiment. If, in a hole in the shutter of a darkened room, is placed a double convex lens, the room is itself a camera-obscura. Let a white screen be placed in front of the lens

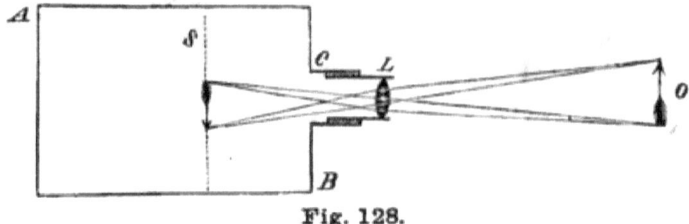

Fig. 128.

at a proper distance, and it is at once covered with a perfect picture of whatever scenery may be outside. Houses and distant hills, the sky with its floating clouds, men and animals in the street, and even the flying birds, and the curling smoke, are distinctly painted in miniature upon the screen.

The Eye. — But most perfect of all optical instruments is the eye. Who could at first believe, that in describing, as we have done, the camera-obscura, we were describing a rough model of the human eye! Yet the eye is a camera-obscura, differing from the common form of that instrument only in its wonderful perfection.

The human eye is a gobular chamber, having for its outer wall a hard tough membrane called the SCLEROTIC COAT. The front part of the sclerotic coat is a transparent substance called the CORNEA. The chamber is lined with a more delicate membrane called the CHOROID, and, to insure the darkness of the place, this is covered upon the inside with a *black paint.* The front part of the choroid coat is called the IRIS, and in the center of this is a round hole called the PUPIL of the eye, through which light may pass into the dark chamber beyond. Behind this opening is a double convex lens, very transparent and considerably hard, called the CRYSTALLINE LENS. Between this lens and the cornea is a limpid liquid called the AQUEOUS HUMOR; and filling the dark chamber, behind the lens, is another fluid, called the VITRE-

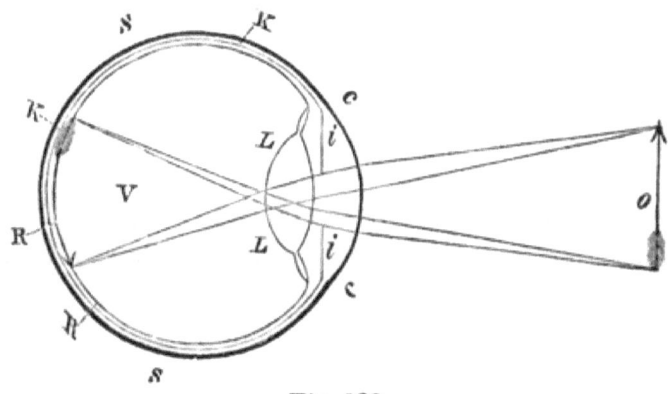

Fig. 129.

OUS HUMOR. The arrangement of these parts may be understood by attentively studying Fig. 129, which represents a section of the eye. S S is the outer or sclerotic coat, sometimes called the white of the eye. *c c* is the cornea; it is more convex than the sclerotic. K K is the choroid, and *i i* is the iris, the vertical curtain which shuts out all light, except what may get through the hole at its center, — the pupil. L L is the crystalline lens, and the large chamber V is filled with the vitreous humor. The course of the rays of light is also shown in the figure. An inverted image of an object, O, is formed at R. It is there received upon a net-work of delicate nerve-fibers called the RETINA, R R.

The mind takes cognizance of this picture, and the person is said to *see the object* O. These pictures on the retina are always smaller than the objects, and, the more distant the object, the more minute the image. The diameter of the eye is little more than an inch; and yet when a person sees an extended landscape, every visible object, far and near, is painted upon the inner lining. If the picture in the human eye be thus minute, what must it be in the eye of a canary-bird or butterfly!

SECTION VI.

ON DOUBLE REFRACTION AND POLARIZATION.

88. When a beam of light passes through a crystal of Iceland spar, it is doubly refracted. The two beams which emerge are both polarized. Light may be also polarized by reflection. The effects of polarized light are numerous and important. (*G*. 636, 642, 645, 646, 655, 658, 661.)

Double Refraction. — A crystal of Iceland spar is very transparent, and its form (Fig. 130) is as regular as could be cut by the hand of a skillful artist. Each of its

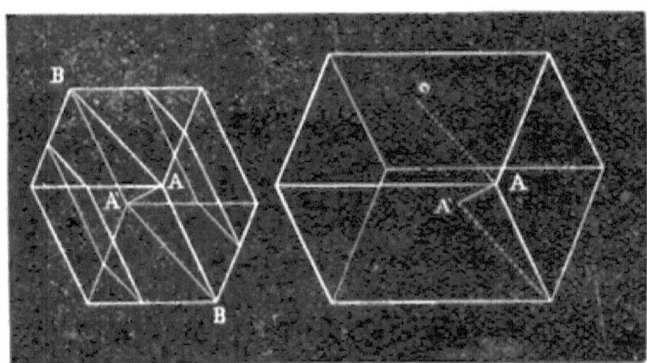

Fig. 130.

six surfaces is a parallelogram. They are so arranged that three of them have each an obtuse angle at A, and the other three each an obtuse angle at the opposite corner, at A′

in the crystal B in the figure. A line joining the points A and A' is called the OPTIC AXIS of the crystal. If the edges of the crystal are all equal the axis is a diagonal, but if not then the axis is not a diagonal of the crystal. Both cases are shown in the cut. Now, if a ray of light be passed through such a crystal in any direction *not parallel to the axis*, it will emerge as two separate rays, and the light will be said to be *doubly refracted*.

Double refraction causes the curious effect of making any thing on which the crystal rests appear to be double. (Fig. 131.) One of these refracted beams obeys the regular law of refraction; the other does not. The first is called the

Fig. 131.

ORDINARY beam, the other the EXTRAORDINARY beam. Many other transparent crystals have this power of double refraction.

Both Beams are polarized. — A very curious change is wrought in the light by double refraction. Common light will pass through any transparent medium, no matter in what position it may be held; but these doubly refracted rays are able to pass through a second medium when it is held in certain positions only. For example, if the ordinary beam be made to fall upon a flat plate of tourmaline (a transparent mineral crystal), and it go through when in one position, it will not go through when the plate has been turned 90°

around. Turn the plate 90° more, and the beam will again pass through it; turn it 90° farther yet, and the beam will be again wholly cut off.

If the extraordinary beam be tried, it will be wholly transmitted by the plate in positions where the ordinary beam was cut off, and wholly cut off where the other was transmitted.

When light, by being refracted or reflected, is made incapable of being again refracted or reflected except in certain directions, it is said to be *polarized*.

Polarization by Reflection. — If a beam of light, shown by a b (Fig. 132), falls upon a plate of black glass at an angle of incidence $54\frac{1}{2}°$, a part of it will pass into the glass, the rest of it will be reflected. If the reflected part be examined by a plate of tourmaline, it will be found to be polarized. Or if another plate of black glass, N, is placed parallel to the first, the beam will be reflected as the figure shows it; but let the plate be turned 90°, as shown by the dotted lines, and the beam will be wholly cut off. Turn it 90° farther, and the reflected beam appears again; another 90°, and it is again cut off. At any other angle of incidence than $54\frac{1}{2}°$, the light will be only partly polarized: $54\frac{1}{2}°$ is the *polarizing* angle for glass.

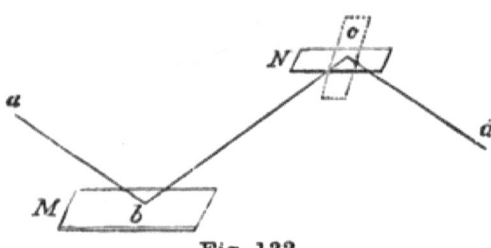

Fig. 132.

Polarizing Instruments. — The instruments, called polariscopes, by which to study polarized light, consist essentially of two parts, one to polarize the light, the other to examine it after it has been polarized. The first is called the POLARIZER, the second the ANALYZER. One of the simplest forms of the instrument is shown in the figure (Fig. 133). The polarizer, P, is a plate of glass, covered on the back of it with black varnish. The analyzer, A, is a plate of tourmaline set into a movable tube. Objects to be examined by polarized light are supported in a movable ring O.

Theory of Polarization. — To explain the phenomena of polarization, we must remember that light consists

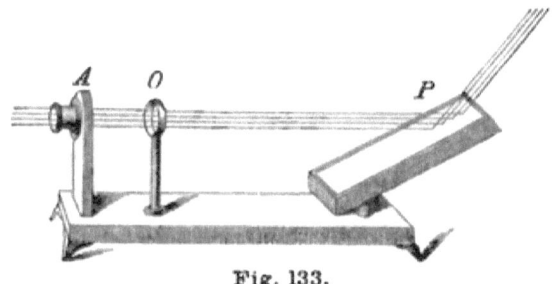

Fig. 133.

of undulations, and add to this the assumption that the vibrations of the ether-particles take place in all possible directions *at right angles* to the *direction in which the ray itself is going*. Let us for a moment suppose that we could see the ether, and that we look squarely at the end of a beam of light. We may fancy that we should see a circular outline, with the particles of ether moving swiftly in the directions of all its diameters. Let A (Fig. 134) represent this view. Now, the theory assumes that by refraction or

Fig. 134.

reflection, all these vibrations are changed into two sets, one in a horizontal plane, B (Fig. 134), and the other in a vertical plane, C (Fig. 134). This change is what is called POLARIZATION. The tourmaline plate, or the plate of glass, will let one of these sets of vibrations pass through it only when in certain positions, owing to some peculiar arrangement of its molecules, but when in position to cut off one set it allows the other to pass freely.

Effects of Polarization. — When a thin plate of mica, or other doubly refracting medium, is put at O, in the polariscope (Fig. 133), the two beams emerging, by interfering, produce most beautiful colors. When seen in certain directions, colored rings of surprising beauty, with a black cross, appear. The form and arrangement of these rings differ in different crystals — a fact of much interest to the mineralogist.

Some substances have the power to change the position of the plane of vibration, in a ray of polarized light. Thus if the analyzer (Fig. 133) is turned so that the polarized light is turned off, a thin plate of quartz at O will cause the ray to re-appear. In this case suppose the vibrations to be in the vertical plane, and that the analyzer is turned to the right just 10°; the quartz must bend the plane of vibration 10° to the right also, in order that the ray may pass. This is called ROTARY polarization.

A great number of liquids have this power. Some of them turn the plane to the right; such is a solution of canesugar: others turn the plane to the left; such is a solution of grape-sugar. This fact is of great interest to the chemist, and it assists the physician at times to determine the healthy or diseased condition of the fluids of the human system.

SECTION VII.

REVIEW.

I. — SUMMARY OF PRINCIPLES.

The undulations of the ether, expending their energy upon the eye and the sense of touch, are recognized as light and heat.

These undulations are transmitted through all bodies with greater or less facility.

But when they strike the surface of a second medium only a part of them enters; another part is thrown back, or reflected, making the angle of reflection equal to the angle of incidence.

All effects of mirrors are explained by this principle of reflection.

The waves which enter the second medium are partly transmitted and partly absorbed.

Those which are transmitted are bent, or refracted, making the index of refraction always the same for the same media, but different for different media.

The effects of lenses are explained by this principle of refraction.

Those waves which are absorbed in their passage through a body are transformed into heat or actinism, expending their energy in warming the body or in causing some chemical change.

In the sunbeam there are a multitude of waves of different rates. The prism is able to separate them, and when separated they affect the eye as different colors. Color, therefore, depends on the rate of the undulation, and is analogous to the pitch of sound.

The intensity of light depends on the amplitude of the undulations; it is analogous to the loudness of sound.

All the phenomena of color are explained on this principle of decomposition of light.

The relation between wave-length, velocity in space, and rate of the undulation, is shown by the formula $L = \dfrac{V}{R}$.

For light, the value of V is found by observations on the eclipses of one of Jupiter's satellites, and also by experiment. The value of L is found by experiment with Newton's rings and in other ways. The value of R may then be found by the formula.

Velocity of light $= 186,000$ miles a second.

Wave-length for red $= .0000266$, and for violet $= .0000167$ inch.

Rate for red = 392 millions of millions, and for violet = 754 millions of millions a second.

Undulations whose rates are less than that of the red are recognized only as heat, and those with rates above that of the violet are recognized only by chemical effects.

Light and radiant heat may be polarized, but sound can not. This difference between light and sound is explained by supposing that the *vibrations* in the wave of light or heat are *transverse*, while in the wave of sound they are *longitudinal*.

II. — SUMMARY OF TOPICS.

68. Wave-front. — Rays of light. — Are transmitted. — Light moves in straight lines. — With uniform velocity. — The law of intensity. — Photometry.

69. Reflection. — The law. — Illustrations. — Vision.

70. Mirrors. — Effect of plane mirrors. — Of concave mirrors. — Foci. — Effect of convex mirrors.

71. Images by reflection. — Image of a point. — Image by a plane mirror.

72. Images formed by a concave mirror. — Images of points. — Of an object beyond the center. — Of an object between the center and focus. — Of an object between the focus and the mirror.

73. Images by a convex mirror.

74. Reflection from rough surfaces.

75. Experiment showing refraction. — Definition. — The first law. — The second law. — Illustrations of refraction.

76. The index of refraction. — The law. — Index varies with the media.

77. Lenses. — Effect of convex lenses. — Foci. — Effect of concave lenses.

78. Images are formed by lenses. — Explanation.

79. Image of an object twice the focal distance. — Of the object farther away. — Of the object at less distance. — Of the object between the focus and the lens.

80. Image by concave lens.

81. Prisms. — Refract light. — Also decompose light. — Recombination of the colors.

82. A pure spectrum. — The spectroscope. — Fraunhofer's lines. — The bright lines. — Spectrum analysis. — Invisible parts of the spectrum, heat. — Another experiment, chemism. — The energy of the sunbeam.

83. The equal diffusion of heat. — Transmission of heat. — Depends on temperature. — Nature. — Condition of surface.

84. The primary rainbow. — Red on the outside. — Colors in form of an arch. — The secondary bow.

85. The color of bodies. — The color of the sky. — The color of the clouds.

86. Sound and light alike. — Interference of sound. — Of light. — The conditions. — Wave-lengths of light. — The measurement. — Color depends on rapidity of vibration. — Diffraction. — Explanation. — Illustrations.

87. The microscope. — Compound. — The telescope, Galileo's. — The astronomical. — The terrestrial. — Reflectors. — The Herschellian. — The magic-lantern. — The camera-obscura. — The eye.

88. Double refraction. — Both beams are polarized. — Polarizing by reflection. — Polariscopes. — Theory of polarization. — Effects.

CHAPTER VIII.

ON ELECTRICAL ENERGY.

SECTION I.

ON FRICTIONAL ELECTRICITY.

89. Electricity may be produced by friction. The electrical machine is an apparatus for this purpose. It may be detected by instruments called electroscopes, showing its action in two ways, — by attraction and repulsion. Its intensity may be measured by instruments called electrometers. It is governed by two laws: —

1st, Electricities of the same kind repel each other, of different kinds attract.

2d, The force of the attraction or repulsion is inversely as the square of the distance between them. (*G.* 704, 711, 714; *A.* 928–931.)

Electricity produced by Friction. — If a well-dried glass tube be thoroughly rubbed with a flannel cloth, it will be found to have new and curious properties. Hold it near the face, and a feeling will be experienced as if a gentle breeze were blowing against the cheek; bring it nearer, and perhaps a prickling sensation will be felt, and it may be that a crackling sound will at the same time be heard; or approach it toward some very light substances, such as delicate bits of loose cotton, and they will rush toward it (see Fig. 135), and remain for a little time clinging to it. These various effects show the presence of electricity: the friction of the flannel upon the glass has produced it.

NATURAL PHILOSOPHY. 225

The Electrical Machine. — The electrical machine is an apparatus for producing electricity by friction. It is represented in Fig. 136. Its principal parts are, 1st, a body

Fig. 135.

upon whose surface electricity is to be evolved; 2d, the rubber by the friction of which electricity is produced; and, 3d, the conductor on which the electricity may be accumulated. In the form shown by the figure, the first of these parts consists of a thick glass plate P, to be turned by a crank. The rubber R is made of leather, covered with an amalgam made of mercury, tin, and zinc. Two such pieces of leather are pressed, one against each side of the plate, by means of a brass clamp, which is supported upon a glass pillar. The conductor, or as usually called the *prime conductor*, C, is a brass ball, or a cylinder with rounded ends, mounted on a glass support. Connected with the prime conductor, is a brass fork F, one prong of which is on each side of the plate, with many sharp projecting points reaching toward the glass.

Its Action. — By turning the crank, the friction of the rubber upon the plate evolves electricity, which remains upon the surface of the glass until it is brought around to the fork, and the prime conductor is thrown into the same electrical condition. We shall soon see how this action is explained on the principle of induction. The glass support prevents the electricity from leaving the conductor. When the machine is in operation the rubber is connected with the floor by a chain. All parts of the machine must be free from dust and thoroughly dry.

When a machine of this kind, of medium size, is in successful operation, the effects of the glass tube are experi-

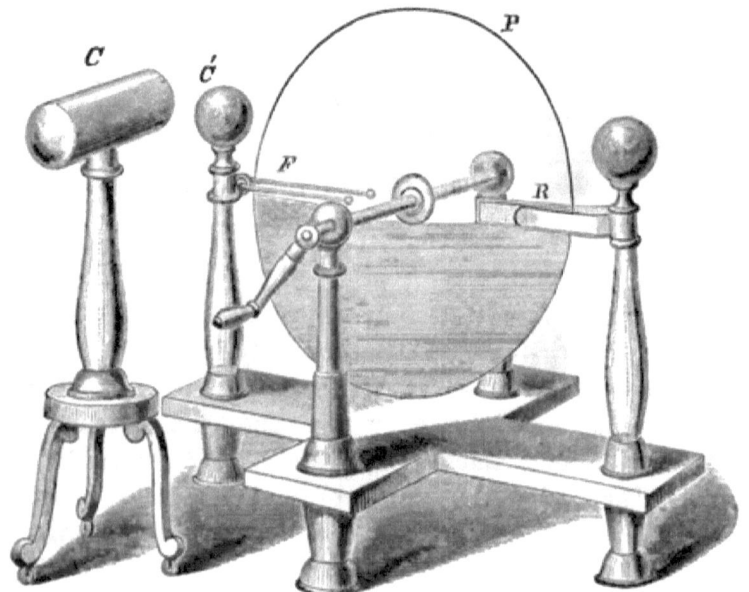

Fig. 136.

enced in a far greater degree. The face or the back of the hand will feel the breezy or prickling sensation at a distance of several inches from the conductor; all light bodies held near it immediately fly to its surface; and, if the knuckle or a brass ball be brought near, bright and zigzag sparks may

NATURAL PHILOSOPHY. 227

be drawn through a distance of from one to two inches, the light being accompanied by a sharp report.

Electricity detected by Electroscopes. — When the electricity is feeble, there should be some more convenient way of showing its presence. Any instrument for this purpose is called an ELECTROSCOPE. The simplest form is called the *pith-ball* electroscope. It consists (see Fig. 137) of a ball of pith from the corn-stalk, or elder, hung by

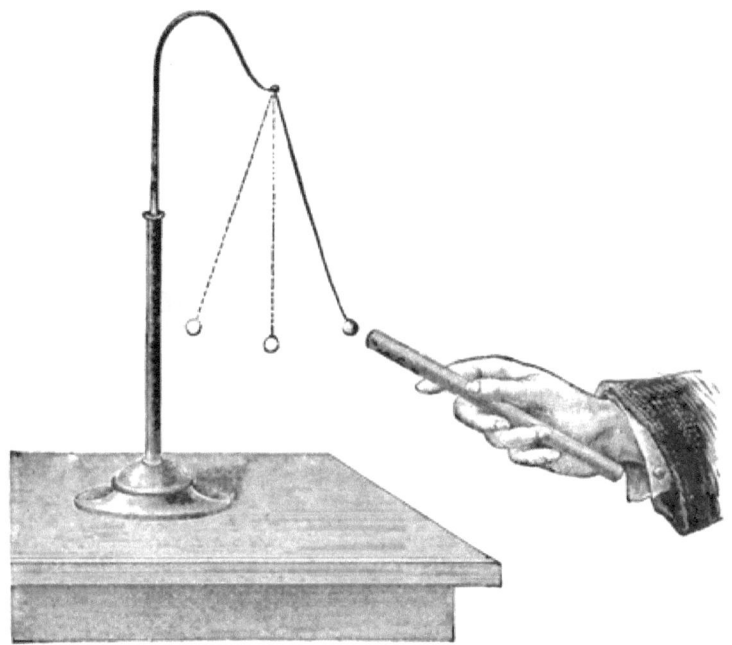

Fig. 137.

a slender silk thread from a glass support. This little ball will instantly announce the presence of electricity by moving toward the body which contains it, and after a moment leaping away again.

Two Opposite Actions. — In this experiment the electricity shows its presence both by attraction and repulsion. If the pith-ball of the electroscope be brought near to the prime conductor of the electrical machine, it will fly toward it, but, on coming in contact with it, will as instantly leap away again.

Now, rub a glass rod with flannel, and hold it near the pith-ball which has been repelled by the conductor; the glass rod will also repel it; but, if a stick of sealing-wax be used in place of the glass tube, the pith-ball will be strongly attracted. Notice, that the pith-ball is *repelled* by the electricity of *glass*, and *attracted* by the electricity of *sealing-wax*. It is thus seen that the electricities of glass and sealing-wax are not alike. To distinguish them from each other, that which is produced on glass by the friction of flannel is called *positive* (+) electricity; that produced upon sealing-wax is called *negative* (−) electricity.

It is found to be impossible to develop one of these conditions without the other also. The positive always appears on one of the bodies rubbed together, and the negative upon the other.

The Law. — We have seen that positive electricity is produced by friction on glass, and that the opposite force is evolved by friction on sealing-wax. Now let two pith-balls be suspended by silk threads so as to be in contact. Thoroughly rub the glass tube; bring it in contact with the balls; they both receive positive electricity from the tube, and it will be found that they will no longer remain in contact. We learn from this experiment that two bodies with the *same kind of electricity repel each other*.

Again: let the sealing-wax be thoroughly rubbed and brought near to the two pith-balls while they are repelling each other, and they will both fly toward it. We learn from this experiment that bodies with *different kinds of electricity attract each other*.

Bodies in the same electrical condition repel one another; in opposite conditions they attract.

Application. — This law furnishes an easy test by which to find out which kind of electric force is, in any case, produced. Is the prime conductor of the electrical machine positive or negative? To decide the question, rub the glass tube; bring it in contact with the pith-ball of the electroscope; the electricity of the ball is thus known to be positive. Now, bring it near the prime conductor of the machine in operation; it is *repelled*. The electricity of the conductor is positive. The electricity of the rubber is negative, because, if the chain be removed, and the electrified pith-ball be brought near the brass mounting of the rubber, it will be attracted.

90. A charged or electrified body, acting through a non-conductor upon an insulated conductor, polarizes it. This action is called induction. (*G.* 722-726, 728.)

A Charged Body. — Whenever by friction, electricity is developed upon the surface of a body, the body is said to be electrified; and if, by bringing another body in contact with it, electricity is imparted, the body which receives it is said to be charged. Thus the glass tube, when rubbed, becomes *electrified;* the pith-ball of the electroscope, coming in contact with the glass, takes electricity from it, and becomes *charged*.

A Non-Conductor. — Some bodies allow electricity to pass freely over their surfaces; such bodies are called Conductors: others will not allow electricity to pass freely over them; these are called Non-Conductors. If a brass rod be held in contact with the prime conductor of a machine, it will be found impossible to charge it; a glass rod held in the same way will not prevent the charge from accumulating. The brass allows the electricity to pass into the person; the glass does not: brass is a conductor; glass is a non-conductor. The metals, as a class, are good conductors. Beside glass, we notice silk, India rubber, and dry air, as being among the best non-conductors.

Potential. — The electric condition of the earth is the standard with which to compare the electric conditions of other bodies. In the earth we suppose the + and − electricities to be just balanced: the electric power is zero. The term *potential* is used in comparing electric conditions. When the condition is the same as that of the earth, the potential is zero.

High and Low Potential. — When the electricity at any place is greatly in excess of that in the earth, it is described as a *high potential*. A less and less excess is spoken of as a lower and lower potential. And, when the electricity falls below that of the earth, the *potential is negative*.

The potential of a body is the difference between its electric power and that of the earth in its neighborhood.

Electro-Motive Force. — Now remember that electricity always passes from a place of high to one of lower potential, as surely as water will run from a high to a lower level. And just as we speak of the force of gravity as the cause of the flow of water, so we speak of *electro-motive force* as *that which urges electricity along over a conductor*.

Electro-motive force may be also described as the difference in the potentials of two places.

An insulated Body. — Whenever a body is quite surrounded by non-conductors, it is said to be insulated. The conductor of the machine is insulated by resting upon a glass support. A body which is not insulated can not be charged.

A charged Body polarizes an insulated Conductor. — A body is said to be polarized when the two opposite electricities both exist upon its surface. To illustrate this important condition, let an insulated metallic ball be connected with the prime conductor of the electrical machine, and let a small insulated conductor be placed near it (see Fig. 138). When the ball is charged, the motion of the pith-balls fastened to the small conductor shows that it is also electrified, and, if its electricity be tested, it will be found to be positive at one

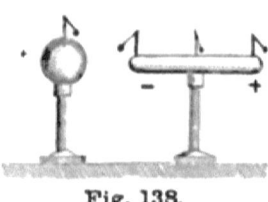
Fig. 138.

end and negative at the other. Both electricities are developed upon its surface at the same time, and the body is said to be polarized. The action of the ball, by which this body is polarized, is called INDUCTION.

If we examine the condition of the polarized body more carefully, we find that in the end next to the ball there is *negative* electricity, and in the distant end there is *positive* electricity. This is always true: when a body is electrified by induction, *the end or side nearest the charged body is always in a condition opposite to that which develops it.*

The insulated conductor is electrified only when near to the ball. Let it be moved away, and the pith-balls drop.

To charge it. — But if, when the conductor is polarized, we touch it with the finger, the electricity which is like that of the charged body will pass off, and the entire surface will remain charged with the opposite kind. It will remain charged, even when taken beyond the influence of the body which polarized it.

91. A series of insulated conductors, placed end to end, near each other, may be all polarized by bringing a charged body near to one of them. Faraday's theory explains induction by supposing the molecules of a body to be polarized, one by another, in the same way. (*G.* 724, 726, 729.)

A Series of Conductors may be polarized. — Let a number of small insulated conductors be placed end to end, near together, with one end of the first one near to a brass ball connected with the prime conductor of the machine (see Fig. 139). The motion of the pith-balls will show that they are all polarized. The effect will be greater if another brass ball, connected with the rubber of the machine, is placed at the other end of the series. The positive and negative electricities are on opposite ends

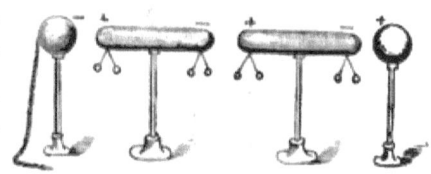

Fig. 139.

of each conductor. All the ends toward the positive ball are negative; all the ends in the other direction are positive.

Faraday's Theory of Induction. — Now, the molecules of one of these conductors are as truly separate bodies as the conductors themselves; and, as one electrified conductor may polarize another, so one of these molecules, acting through the minute distance between them, may polarize another. This polarizing influence passes from one molecule to another, until all the molecules of the body are thrown into this condition, each molecule having opposite electricities on its opposite sides.

Difference between Conductors and Non-Conductors. — The theory goes further, and supposes that the molecules of conductors discharge their electricity easily into one another, while those of non-conductors do not. For this reason, the molecules of the air between the charged ball and the end of the conductor are polarized, and *retain* their electricities, while the molecules of the conductors, as fast as they are polarized, give their electricity to their neighbors. The positive electricity given from one to another, in one direction, accumulates at one end of the conductor; the negative, given from one to another in the other direction, accumulates at the other end.

Polarization precedes Electrical Attraction. — The first action of an electrified body is to *polarize every other* in its neighborhood. The attraction of pith-balls or cotton (Figs. 137 and 135) by the excited glass takes place *after* they have been polarized.

Illustration. — Fig. 140 shows a chime of bells, which are to be rung by electricity. The two outside bells are fastened by metal chains to a rod of metal which hangs from the end of the

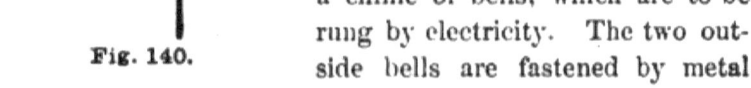

Fig. 140.

prime conductor of an electrical machine. The middle bell is hung by a silk thread, and has a chain passing from it to the floor. Finally, notice two little balls of metal between the bells; these balls are hung by silk threads also. When the machine is in operation, these little balls will fly back and forth and ring the bells merrily.

Now when the outside bells are +, they polarize the balls. The sides of the balls nearest the bells are —, and hence bells and balls *attract*. The balls then strike the bells, become + by contact, and balls and bells *repel*. The middle bell is polarized — : the + balls strike it, and discharge their electricity, which passes off into the earth by the chain. This series of actions is repeated over and over again. Polarization, attraction, charge, repulsion, discharge, follow repeatedly in regular order.

In the Electrical Machine. — We are now prepared to see how the prime conductor of the electrical machine becomes charged. The glass plate near the fork is electrified with + electricity. It polarizes the prime conductor near it. The — electricity which accumulates in the adjacent end is discharged from the points of the fork upon the surface of the glass, and this leaves the conductor charged with + electricity.

92. The Holtz machine is an instrument with which to develop electricity by the continuous inductive action of an electrified body. (*G.* 736.)

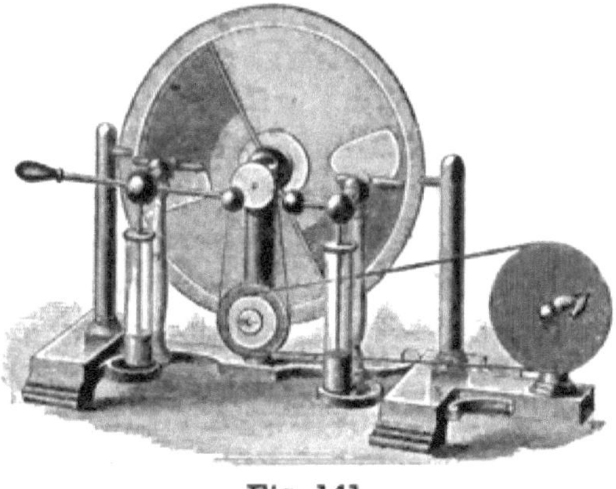

Description. —

Fig. 141.

One form of the Holtz machine is shown in Fig. 141.

Two thin glass plates are insulated, as near together as

possible without touching. The larger plate is stationary, while the smaller one may be rotated very swiftly by a wheel and pulley. Two windows are cut in the stationary plate, and two paper sectors, called armatures, are cemented against the back side of it, on opposite sides of the windows. From the edge of each sector a set of points project into the window. Opposite these points, and separated from them by the revolving plate, are two brass combs. These combs are also connected each with a brass ball in front of the machine, and these balls are on sliding rods so that the distance between them may be changed at will. Each comb and ball are in connection with the inside of a Leyden-jar, which we shall very soon describe, while the outsides of the jars are joined by a chain.

Its Action. — To put the machine in action, one of the armatures is first electrified by bringing an electrified piece of vulcanite against it, and then turning the wheel. In a few seconds the difference of potential in the two jars becomes so great that they discharge in a series of vivid sparks leaping between the balls in front.

Explanation. — The + armature polarizes the comb in front of it, and attracts the − electricity which is projected from its points upon the surface of the revolving plate, leaving the comb charged with + electricity. The other armature is −, and by a similar induction causes a − charge in the other comb. These + and − charges accumulate in the Leyden-jars until they discharge between the balls. Such, in a general way, is the theory of the Holtz, but the details of its action are complex and puzzling. For a full explanation the student may consult a larger work on physics.

93. The Leyden-jar is an apparatus for accumulating electricity by induction. It may be charged by bringing one of its coatings in contact with a charged body, the other being in contact with conductors. It may be discharged by making a conducting communication between its two coatings.

NATURAL PHILOSOPHY. 235

The Leyden battery consists of several Leyden-jars connected. (*G.* 743, 757, 758; *A.* 945.)

The Leyden-Jar. — The Leyden-jar consists of a glass jar, coated both inside and outside with tin-foil, to within a few inches of the top, and provided with a cover of hard dry wood, through which passes a brass rod, with a ball upon its upper end, and a chain reaching from its lower end to the bottom of the jar.

It will be seen by this description, that in this instrument there are *two conducting surfaces, separated from each other by a non-conductor*.

Various Forms. — This idea may be embodied in a variety of forms, any one of which will act on the principle of the Leyden-jar. Thus a pane of glass, coated with tin-foil on both sides, to within a little distance of the edge all around, has the essential parts of the Leyden-jar. A glass goblet partly full of water, and grasped by the hand, illustrates the same idea: the glass, a non-conductor, separates two conducting surfaces — the water on the inside, and the hand upon the outside.

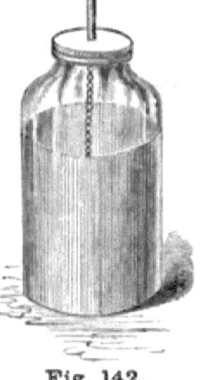

Fig. 142.

It may be charged. — By bringing the ball of the Leyden-jar in contact with the prime conductor of the machine, positive electricity passes into the inside coating. This positive electricity polarizes the glass and the outside coating, causing its surface next the glass to be negative, and the other to be positive. If in contact with a conductor, this positive electricity will pass off, and thus leave the outside coating permanently charged with negative electricity. When by this action the two coatings have opposite electricities, the jar is said to be *charged*. It may be removed from the prime conductor, and remain charged, because the two electricities attract each other without a chance for discharge through the glass. The jar may be handled without danger, if care be taken not to touch the ball and the outside at the same time.

The jar is charged with positive electricity when the *inside* is positive: it is charged with negative electricity when the *inside* is negative.

It may be discharged. — The difference of potential between the inside and outside is very great, and, when a conducting communication is made between the two coatings of the jar, the electricity passes until equilibrium is restored, and the jar is said to be discharged. The conducting communication may be made in many ways. The *discharger* is a convenient instrument for the purpose. It consists of two bent brass arms, with a ball upon one end of each, the other ends being fastened by a joint to a glass handle. Taking hold of the glass handle, bring one ball in contact with the outside of the jar, and the other near to the knob; a bright spark and a sudden report announce the discharge.

The coated glass plate and the goblet of water, mentioned before, may be charged and discharged in the same way as a Leyden-jar. To charge the goblet, for example, let a chain from the prime conductor of the machine hang into the water; grasp the outside of the glass while the machine is in operation. Positive electricity will be given to the water; negative electricity will be induced upon the hand, and the goblet is thus charged. Now with the other hand try to remove the chain: the moment the chain is touched, a slight *shock* will be felt, announcing the discharge which occurs.

The Leyden Battery. — The larger the surface of the coatings of the jar, the more powerful will be the charge accumulated. We can obtain a larger surface by using a larger jar, or it may be done by taking several small ones and joining their surfaces by conductors. In the last case, the Leyden battery will be formed. When the inside surfaces are all connected by conductors reaching from knob to knob, and the outsides all joined by standing the jars on a metallic surface, the battery may be charged and discharged as a single jar. It is equivalent to a single jar large enough to have the same extent of surface.

94. The electricity of the atmosphere is of the same nature as that produced by friction. Lightning is the discharge of oppositely charged clouds, illustrating, on a grand scale, the action of a Leyden-jar.

Electricity of the Atmosphere.—The atmosphere is very generally in an electrified condition. This may be shown by raising a metallic rod to a considerable height above the ground, having an electroscope fastened to its lower end, which should be insulated. A *sensitive* electroscope will usually indicate positive electricity, its intensity increasing as the air from which it is drawn is higher. In its ordinary state, the electricity of the atmosphere is always positive: stronger in winter than in summer, and during the day than the night. In cloudy weather the electrical state is uncertain, sometimes changing from positive to negative and back again in a few minutes. On the approach of a thunderstorm these changes follow each other, at times, with remarkable swiftness.

It is of the same Nature as frictional Electricity.—The bright flash and loud report which announce the discharge of a Leyden-jar or battery can not have failed to remind one who has observed them, of the brighter flash and louder report of atmospheric lightning and thunder. These grand and sometimes awful displays of electricity are caused by the same agent which, produced on a glass tube, lightly pricks the cheek or attracts a pith-ball.

To Dr. Franklin belongs the immortal honor of proving the identity of electricity and lightning. A kite was the simple instrument which he employed. Having made a kite by stretching a silk handkerchief over two sticks in the form of a cross, he went out into a field, accompanied only by his son; raised his kite; fastened a key to the lower end of its hempen string; insulated it by fastening it to a post by means of a silk cord, and anxiously awaited the approaching storm. A dense cloud, apparently charged with lightning, soon passed over the spot where he stood, without causing his apparatus to give any sign of electricity. He was about to give up in despair, when he caught sight of some loose fibers of the hempen cord bristling up as if repelled. He immediately presented his knuckle to the key, and received an electric spark. The string of his kite soon became wet with the falling rain; it was then a better conductor, and he was able to obtain an abundance of sparks from the key. By this experiment he furnished a decisive proof of the identity of lightning and electricity.

Lightning is the Discharge of oppositely Charged Clouds. — Clouds are often charged with electricity. When two of them, with opposite kinds of electricity, come near enough together, they will act like the two charged coatings of the Leyden-jar, the air between them being a non-conductor like the glass. When the charge rises high enough, a discharge takes place; the spark of the discharge being a flash of lightning, and its report a thunder-peal. Considering the large extent of cloud surfaces discharged, we need not be surprised at the magnitude of the spark, nor at the deep intensity of the sound.

When the discharge is not hidden by clouds, we can trace the whole length of the spark, and we witness *chain-lightning:* but at other times the spark is behind the clouds; we see only the light of the discharge spread over the surface of the clouds, and this gives rise to what is called *sheet-lightning.*

At times the earth and a cloud are the two charged surfaces, and a discharge takes place between them. Such discharges are the source of danger to life and property. Animals, trees, buildings, all these are better conductors than air, and electricity always chooses the best conductors in its passage. In going from a cloud to the earth it takes these bodies in its way; animals are often killed, trees shattered, and buildings torn to pieces or set on fire.

95. A body having points projecting from its surface can not be charged even when insulated. Or, if a pointed conductor be held toward its surface, it will prevent a charge from accumulating. Upon this principle, buildings are protected from the effects of lightning by lightning-rods.

The effect of Points. — It is found to be impossible to charge a conductor when there are sharp points on its surface, or held near to it. Fasten a pointed wire to the prime conductor of the electrical machine, and the sparks, which before could be drawn from it in abundance, cease altogether, and even pith-balls fail to detect the presence of the force. Or take the pointed wire in the hand, and present its point to the prime conductor, within a few inches of its surface; not a spark can be drawn from it, nor will the pith-balls show either attraction or repulsion. The discharge is silently effected by the air in front of the points. Its molecules become polarized, and are first attracted to the point and then repelled. On coming in contact with the point, they take electricity from it, and move away: others being polarized are attracted, receive electricity, and pass away. Thus the electricity of the body is silently carried off from the point. That such currents of air do really exist, may be proved by various experiments. If, for example, the check, or the back of the hand, be held near to the point, the breeze will be felt; or, if the small flame of a lighted taper be held just in front of the point on the prime conductor, it will be blown away from it, and may even be extinguished.

Lightning-Rods. — We have seen that because buildings are better conductors of electricity than air, they are liable to injury from strokes of lightning. But, since pointed conductors silently discharge the force from a charged body, why not disarm the cloud of its lightning by the use of pointed metallic rods? This question was no sooner suggested to the practical mind of Franklin, than a trial was made, which verified his bold conjecture.

Conductors for the purpose of protecting buildings from the effect of lightning are called lightning-rods. They should be made of metallic rods, pointed at the upper end, reaching several feet above the highest part of the building which they are designed to protect, and downward, without interruption, into the ground below its foundation, far enough to be always in moist earth.

96. The effects of frictional electricity are mechanical, chemical, and physiological. (*G*. 760–762, 770; *A*. 949, 951.)

Mechanical Effects. — We have already had abundant illustrations of *motions* caused by electricity. Poor conductors are also pierced or torn by the electric discharge. To illustrate this by experiment, let the charge of a Leyden-jar be passed through a piece of cardboard; the card will be pierced with a burned or ragged perforation. This effect is produced on a large scale by the lightning-stroke; even rocks are sometimes shattered, while trees are often splintered from top to root, and their fragments scattered far and near in all directions.

Chemical Effects. — The chemical effects of electricity are shown through the agency of the heat which it develops. To illustrate by experiment: wrap the ball of a Leyden-jar with loose cotton, and sprinkle upon this, very finely powdered resin. This done, charge the jar powerfully, and then discharge it by bringing first one ball of the discharger in contact with the outside of the jar, and then the other a little above its hooded knob. The discharge takes place through the resin, and sets it on fire. Buildings are sometimes set on fire by the lightning-stroke.

Physiological Effects. — The effect of electricity upon

the human system is peculiar and startling. No description can give a correct idea of it: it must be experienced by one who would know what it is. Let a person place one hand upon the outside surface of a lightly charged Leyden-jar, and with the other hand touch its knob. He will find that he can no longer control his muscles: his hands are, on the instant, suddenly jerked, while a peculiar and almost indescribable sensation is felt in the wrists and arms.

Many persons by joining hands may form an unbroken connection between the two coatings of the jar, and all at once experience these effects.

SECTION II.

ON MAGNETIC ELECTRICITY.

97. Magnets are either natural or artificial, and may be made in different forms; but in any form the magnetism is stronger at the ends than in the middle. The ends are called poles. (*A.* 967, 974; *G.* 670, 895.)

Magnets. — Bodies that attract iron in preference to other metals are called MAGNETS.

Fragments of an ore of iron are sometimes found, which have the properties of a magnet. Such a fragment is a NATURAL MAGNET or LODESTONE.

If a bar of iron or steel be rubbed against a magnet, it will become magnetic; it will then be an artificial magnet.

Soft iron or steel will lose its magnetic properties quickly; hardened iron or steel will retain them.

The *bar magnet* is a straight bar of steel; the *horseshoe magnet* is a magnet whose shape is that of a horseshoe, or the letter U (see Fig. 143); its ends are thus brought near together. A piece of soft iron across the ends, N S, is called the *armature*.

Experiment. — To illustrate their peculiar preference for iron, let some iron-filings be mixed with some filings of brass; bring one end of the magnet among the filings, and on removing it great numbers of the iron particles will be seen clinging to it, while the brass particles are all left behind.

Their Force is Stronger at their Ends. — If a bar magnet be rolled in a bed of iron-filings, large clusters of them will be found clinging to its ends, their numbers getting less toward the middle of the bar, where very few, if any, will be held. (Fig. 144.) By this experiment we learn that the magnetism is not equally distributed over the surfaces of magnets, but, on the contrary, that it is strong at the ends and weak or neutral in the middle. The ends are called POLES, one being a *north* pole, the other a *south* pole.

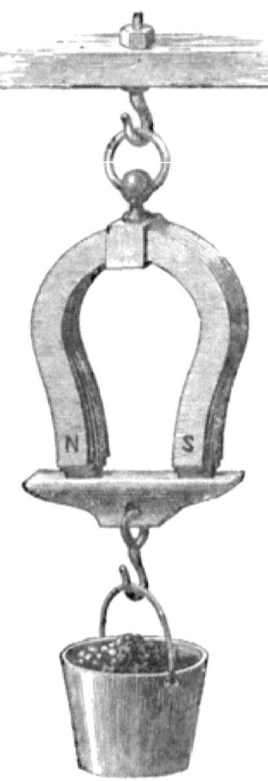

Fig. 143.

98. Magnetism shows itself both by attraction and repulsion, obeying the following law: Poles of like names repel each other; those of different names attract.

Attraction and Repulsion. — Iron which is not magnetized will be attracted equally by both poles of a magnet; it is not so when two magnets act upon each other. By presenting the south pole of a magnet to the north pole of another, it will show an attraction for it, while the north pole, being presented to the north pole, will repel it. Thus magnetism, like electricity by friction, shows itself by both attraction and repulsion.

Fig. 144.

It is also evident from these experiments that *poles of the same name repel, while those of opposite names attract.*

99. A magnet, like a charged body, will polarize a bar of iron brought near to one of its poles, always inducing magnetism of the opposite kind in the end next to it. The polarizing influence may extend through several bars placed end to end.

It is supposed that every molecule of a magnet is in a polarized state, the north polarity being on the same side of them all, and the south polarity on the other side. (*G*. 669.)

A Magnet will polarize a Bar of Iron. — If a bar of iron $n\ s$, and a magnet N S (Fig. 145), be placed end to end, the iron itself becomes a magnet. That it is a magnet, may be shown by its power to attract or repel the poles of another magnet. Both kinds of magnetism are developed in it, and hence we call it polarized.

Each pole of a magnet will always induce the opposite kind of magnetism in that end of the bar which is nearest to it.

Unlike frictional electricity, there is no discharge of magnetism when opposite kinds are brought together: the polarization takes place even when the bar is in contact with the magnet, and, if the bar be made of steel, the polarity remains after the magnet is removed.

Fig. 145.

Several Bars may be polarized. — A second bar may be placed with one end near to the first, and it will be found to be polarized; so a third may be polarized by the second: the series may be continued further, but the force is less in each successive magnet. To illustrate by experiment: from the north pole of a strong bar magnet hang a key; from the lower end of this one a smaller key may be hung; a third still smaller, or a nail, will be held by this, and a tack will cling to the lower end of the last. The series of keys and nails has become a series of magnets, each with its north and south pole, their north poles all directed downward. (See Fig. 146.)

All the Molecules of a Magnet are polarized. — Now, the molecules of a magnet are as truly separate from each other as the several magnets in the series just described, and it is thought that each molecule is a magnet, with its north and its south pole. Acting through the minute distances that separate them, each one is polarizing its neighbors; and hence, like the series of bars, their north poles must be all arranged in one direction, their south poles in the other. Both kinds of magnetism act upon each separate molecule, and keep it in a magnetic state. There is no transfer of the force from one molecule to another, as there is of electricity in a charged body, so there can be no discharge of magnetism. A magnet, like a body charged with electricity, may *polarize* another, but it can not, like the charged body, become neutral by giving up its force.

Fig. 146.

Why, then, is the middle of a magnet neutral, while only toward its ends do the forces show themselves? Not because the force of one kind leaves the molecules of one end and goes to the other, but because in the middle of the series the two forces are equal and in opposite directions, and must neutralize each other.

100. If a bar magnet be supported so as to move freely in a horizontal direction, it will rest only when its poles point north and south or nearly so; its variation is subject to both annual and diurnal changes. (*G.* 679.)

If a Bar Magnet be supported. — A magnet may be supported in three ways so as to have free motion. It may be balanced upon a pivot. (See Fig. 147.) It may be hung from a fixed support by a fine thread tied about its middle point (Fig. 148). Or it may be, for purposes of simple experiment, fastened to a cork, and laid upon water.

It will point North and South. — The magnet supported in either of the ways mentioned will swing back and forth until it finally settles to rest, and it will then be found pointing north and south. The end which points north is called the NORTH POLE: it is evident, however, that its magnetism is of the kind opposite to that of the north magnetic pole of the earth toward which it points.

A slender bar magnet thus balanced is called a MAGNETIC NEEDLE. Such a needle is used by mariners to direct them in their long voyages across the ocean. For this purpose it is placed over a card upon which the "points of compass," north, south, east, west, and others, are marked; and, for protection, put into a box supported by pivots, so that it will keep the needle in a horizontal position amid all the rolling or plunging motions of the ship. Such an arrangement is called the MARINER's COMPASS.

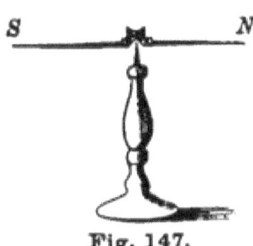
Fig. 147.

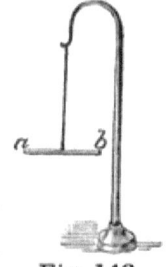
Fig. 148.

Its Variation. — While it is true that the needle points in a direction which may be described as north and south, we must not understand that this description is exact. Indeed, the needle seldom points *exactly* north and south. There are places at which it does; there are others at which it points east of north; and others at which it points west of the true north and south line. Its deviation, or, in other words, what it lacks of pointing in a true north and south line, is called its VARIATION.

The Line of no Variation. — If those places on the earth's surface at which the needle points due north and south be joined by a line, this line is called the LINE OF NO VARIATION. This line goes quite around the globe in a north and south direction. It is, however, an irregular line, bending now to the eastward and then to the westward. We may trace its general course through North America, by remembering that it strikes the continent near Cape Lookout, on the coast of North Carolina, passes through Staunton in Virginia, a little east of Cleveland in Ohio, across Lake Erie, and thence onward to Hudson Bay.

At places east of this line the variation is toward the west; at places west of it the variation is toward the east.

The Annual Variation. — The variation of the needle at any place is continually changing. For example: the variation at Washington, D.C., was 36' west in the year 1800, but in 1860 it had increased to 2° 54'. Such a change is going on year by year at all places. The variation increases for several years, and then again diminishes. So the needle vibrates, first westward, then eastward, and back again, taking many years to make a single vibration.

The Daily Variation. — Besides this annual variation, the needle has a daily variation, much greater in summer than in winter — amounting to about 15' in the former, and only about 10' in the latter. At about 8 A.M., the north pole begins to swing westward, and this motion continues until about 1 P.M. Soon after this time it slowly moves back toward the east until, at about 10 P.M., it has reached its starting point. It then moves west again until about 3 A.M., after which it swings back to the eastward until 8 A.M. It completes these two full vibrations every twenty-four hours.

101. If a magnetic needle be allowed to move freely up and down, it will seldom rest in a horizontal position. Its inclination is called the dip of the needle.

In the northern hemisphere the north pole of the needle dips; in the southern hemisphere the south pole dips.

The Dip of the Needle. — If a slender steel needle be balanced upon a horizontal axis so that it may freely move up and down, and *be then magnetized*, it will be no longer balanced: the north pole will sink until the needle takes a position very much inclined (Fig. 149). The amount of this inclination is called the DIP OF THE NEEDLE.

In the southern hemisphere the needle also takes an inclined position, but it is the south pole that points downward.

As the dipping needle is carried farther north, the dip increases until a point is reached where the needle stands in a *vertical* position. Of course this point must be the north *magnetic* pole of the earth: it is a curious fact, that it is not the same as the *geographical* north pole. It is in latitude 70° 5' N. and longitude 96° 45' W., — a little north and west of Hudson Bay. It was found by Capt. Ross, in the year 1832.

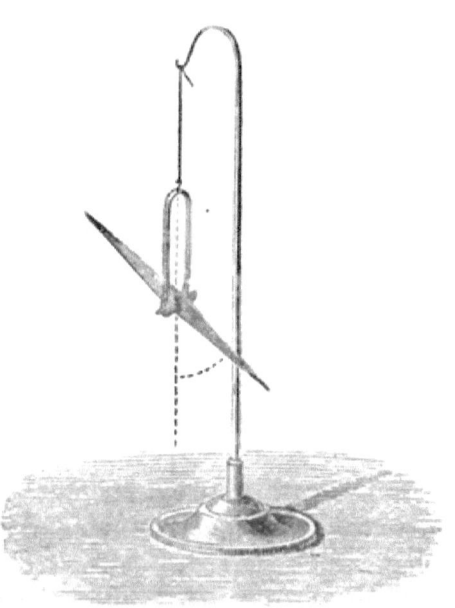

Fig. 149.

Then, traveling southward in the southern hemisphere, the south pole of the needle dips more and more, and there is evidently a south magnetic pole of the earth. This point has never yet been found.

SECTION III.

ON DYNAMICAL ELECTRICITY.

102. A current of electricity may be obtained by chemical action in a simple voltaic cell, or better in a Grove, a Bunsen, or other form of battery. (*G.* 772-774, 783.)

Different Names used. — Frictional electricity shows itself by a single discharge, or, in the case of the sparks of a Holtz machine, in a series of discharges which follow at intervals; but in DYNAMIC electricity there is a continuous action like the steady flow of a stream, and hence it is very properly called CURRENT electricity. Because it was discovered by Galvani it has been called GALVANIC electricity; and, because of Volta's valuable researches, it is also known as VOLTAIC electricity.

Chemical Action. — Put some bits of common zinc into a goblet, and pour upon them some weak sulphuric acid. The fluid will soon begin to boil violently, and bubbles of hydrogen gas will be given off, so that often, if a lighted match is held near, the gas will take fire. This will give the curious appearance of *water on fire*. After a while the action will stop, but not until much and perhaps all the zinc has been used up.

In this case both the zinc and the acid are changed into other substances, and on this account the action is called a CHEMICAL ACTION.

None with Pure Zinc. — Common zinc is impure, and if the pure metal be used for the experiment almost no chemical action will occur. Or let the zinc be *amalgamated*, that is, covered with a coating of mercury, and there will be very little chemical action.

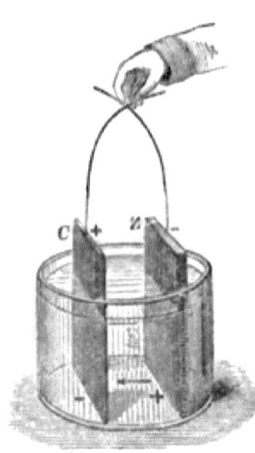

Fig. 150.

No bubbles of gas form where copper is used instead of zinc.

The Simple Cell. — The simple cell is represented in Fig. 150. Into a glass vessel is put a quantity of water, mixed with a little sulphuric acid. A strip of copper and another of amalgamated zinc are inserted in this liquid, and from the upper ends of these strips two metallic wires project. Now, when these two wires are brought together, a multitude of little bubbles of gas rising *alongside of the copper* strip show that there is chemical action; and, if the ends of the wires be brought together and then carefully separated in the dark, a very delicate spark may be seen between them, showing that electricity is produced.

The Chemical Action. — The zinc takes the place of hydrogen in the acid, and forms zinc sulphate. This remains dissolved in the fluid while the hydrogen is forced along from molecule to molecule through the acid until it reaches the copper plate where it escapes in bubbles. This chemical action is believed to be the chief source of the electricity in the circuit.

The electric Conditions in the Cell. — The zinc *in the acid* is *positive*, the copper *negative*. In other words, the surface of the zinc has a *high potential*, the surface of the copper a *low potential*, and hence electricity passes *from zinc through the liquid to copper* to restore the equilibrium. But the chemical action again quickly renews the difference of potential, and another discharge follows. A very rapid charge and discharge throughout the circuit — so rapid as to seem continuous — is what is called the ELECTRIC CURRENT.

The Direction of the Current. — From the + or *generating* plate (which is always the metal acted upon by the fluid) the + electricity passes *through the liquid* to the − plate, then *through the wires* back to the + plate.

The Poles. — The ends of the two wires are called the POLES or the ELECTRODES of the circuit: that from the copper strip is the *positive* pole; the one from the zinc strip is the *negative* pole.

Enfeeblement of the Current. — The current in this simple cell rapidly grows weaker. There are three principal reasons for this; viz.: —

First, The acid is being neutralized by the zinc. The chemical action diminishes on this account. This difficulty can only be remedied by renewing the acid from time to time.

Second, There is local action, that is, an action between the impurities and the particles of the generating plate, by which a multitude of little circuits are made, so that the energy of the plate is not thrown into the general current.

This difficulty can be remedied by amalgamating the zinc, which is done by immersing the clean zinc in mercury. No chemical action then occurs until the wires are brought together, or the circuit is *closed*.

Third, There is the "polarization" of the negative plate. A layer of hydrogen gradually fixes itself upon the surface of the copper, and we have a surface of hydrogen instead of copper. The plate, in this way, gradually becomes + instead of −, and the difference of potential disappears.

This difficulty can be remedied only by managing to use up the hydrogen in some way, so that it shall not come in contact with the − plate. This may be accomplished by using a second fluid.

In Grove's Cell.— In Grove's cell two metals, zinc and platinum, and two liquids, dilute sulphuric acid and nitric acid, are used. The peculiar mode of putting these together may be understood by an attentive study of Fig. 151.

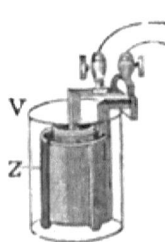
Fig. 151.

A glass vessel, V, is partly filled with dilute sulphuric acid. Into this fluid is placed a zinc cylinder, Z, with a slit from top to bottom, to allow the fluid to circulate, both inside and outside of it, freely. Inside of the zinc cylinder is put a *porous* earthenware cup. Into this cup is poured strong nitric acid, and a strip of platinum is inserted in this fluid. One wire is fastened to the zinc, and another to the platinum. These may be brought together to close the circuit.

The *platinum pole* is *positive;* the zinc pole is negative.

What becomes of the Hydrogen?— There is the same chemical action as in the zinc-copper or simple cell, but the hydrogen, on its way to the platinum, must enter the nitric acid. The nitric acid is decomposed by the hydrogen which takes oxygen from it, and the two remain in the harmless form of water.

Bunsen's Cell. — Bunsen's cell (Fig. 152) differs from the one just described, by having a carbon cylinder, or rod,

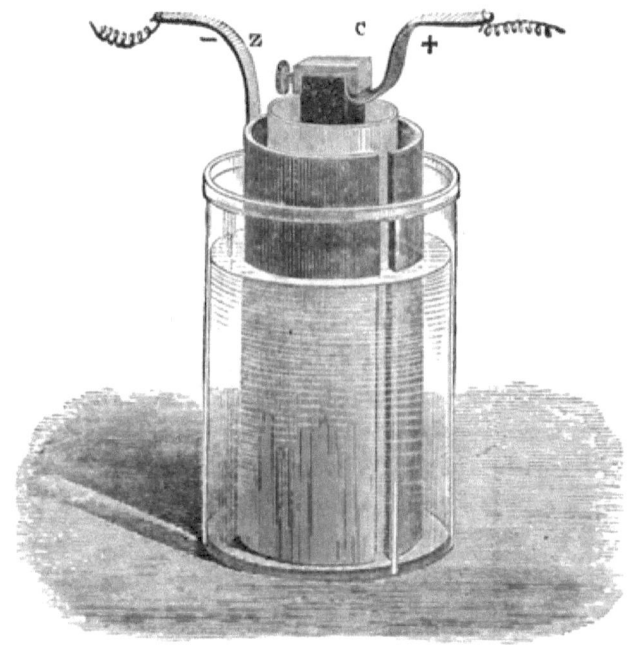

Fig. 152.

in place of the strip of platinum. This does not greatly diminish its action, while it makes it much cheaper, platinum being a very costly metal.

Other Forms. — There are many varieties of voltaic cell in market. They are made by ingeniously varying the materials and shape to fulfil the condition for a current.

This is the Condition. — There shall be two metals and a liquid so related that the liquid shall act chemically upon one more than upon the other.

103. Ohm's law states that the strength of the current is equal to the electro-motive force divided by the resistance.

Definitions. — "The force by which electricity is set in motion in the circuit is called the ELECTRO-MOTIVE FORCE." It is commonly regarded as the difference of potentials maintained by the cell.

"The quantity of electricity which, in a unit of time, flows through a section of the circuit, is called the STRENGTH OF CURRENT." It is also called INTENSITY.

No substance is a perfect conductor of electricity, and some substances forbid its passage altogether. This opposition to the passage of the current is called RESISTANCE.

The resistance which a wire offers to the action of electricity through it depends upon the material of which it is made, and upon its size. The metals are the best conductors, silver standing at the head of the list, copper next, and lead being among the poorest. The larger and the shorter the wire, the less resistance it offers.

The Formula. — The relation of these values as stated above in Ohm's law may be simply expressed in the formula, $C = \dfrac{E}{R}$; in which C stands for strength of current, E for the electro-motive force, and R for the resistance.

104. The resistance in a circuit is partly in the cell, *internal*, and partly in the conductors outside, *external*.

The best effect is obtained in the use of a battery when the internal and external resistances are equal.

If the resistance to be overcome is small, we need a battery of low resistance; but, if it is great, a battery of high resistance.

A Battery of Low Resistance. — The resistance in the cell depends chiefly on the fluid, because the conductivity of the metals is so vastly greater than that of the fluid.

Now, we have seen that the resistance of a conductor

depends on the *area of its cross section*. If we enlarge the plates in the cell we make the area of the conducting liquid between them larger, and so diminish its resistance.

This is most conveniently done by taking several common cells, and joining all their zincs together by a wire outside, and all their coppers by another wire. This forms a *battery of low resistance*. It is sometimes called a battery for *quantity*.

Its Electro-Motive Force. — In the zincs, all joined together, there is the same potential as if there were only one alone. There is no greater *difference* of potential in zincs and coppers than if there were a single cell; that is, the electro-motive force is not increased.

Its Strength of Current. — By linking cells in this way, the *internal* resistance is diminished, while the electro-motive force remains the same. If there were no *external* resistance, then R in the formula $C = \dfrac{E}{R}$ would be less and less as the number of cells increases, and hence $\dfrac{E}{R}$ would be greater; that is, the strength of current, C, would be greater in proportion to the number of cells.

Hence, when in using electricity, there is little *external* resistance, a *low-resistance* battery should be used.

A Battery of High Resistance. — We may link the cells together in another way. Let the zinc of one be joined to the copper of the next, and its zinc with the copper of the next throughout the series, and finally close the circuit by joining the first copper with the last zinc.

This forms a *battery of high resistance*. It is sometimes called a battery for *intensity*.

Its Electro-Motive Force. — In this case each zinc acts separately, and the electro-motive force of the battery is the sum of the electro-motive forces of the separate cells. For n cells of the same kind it would be n E.

The Resistance. — The *internal* resistance is also increased in the same ratio, for a dozen cells yield a dozen times as much resistance as one cell. But the *external* resistance is not at all affected by changing the number of cells; and if we suppose it to be very, *very much* greater than the *internal* resistance of one cell, it would also be much greater than the resistance of a dozen cells, and the *total* resistance may be *very little* greater with n cells than with one.

The Strength of Current. — Then in the formula $C = \dfrac{E}{R}$, with n cells the numerator is increased to n E, while the denominator remains almost unchanged. Hence C is increased almost n times. That is: *The strength of current increases almost in proportion to the number of cells.*

Hence, when in using electricity, there is great *external* resistance, a *high-resistance* battery should be used.

Quantity and Intensity. — Large *surfaces* of elements in the battery yield large *quantities* of electricity: a large number of cells in series yields electricity of great *intensity*.

The greatest difference between current electricity and frictional electricity is this: current electricity is remarkable for its *great quantity* but *feeble intensity*, while frictional electricity is equally remarkable for its *great intensity* but *small quantity*.

105. Current electricity produces: —

Heat whenever it is resisted in its action.

Light whenever its intensity enables it to leap through air, or to render a poor conductor incandescent.

Magnetism whenever its conductor encircles a bar of iron.

Chemical action whenever it goes through an electrolyte. (*G.* 809–813, 816–820, 854, 856.)

Heat. — Electricity, when resisted in it's action, shows itself as heat. When it acts through a fine wire, the wire may be made red-hot, and in many cases melted, by the heat produced. Several inches of fine iron wire may be thus melted by a battery of twelve or fifteen cells. This power of electricity is applied to the exploding of gunpowder, for blasting rocks. For this purpose, a cartridge is made by filling a tin tube with gunpowder, and corking its ends tightly. Through one of the corks two copper wires pass, joined in the powder by a fine steel wire soldered to their ends. The copper wires are then connected with the poles of a distant battery. The instant that the circuit is made, the fine wire in the gunpowder becomes intensely hot, and its heat explodes the gunpowder.

Light. — When the wires which lead from the poles of a powerful battery are tipped with charcoal points, if these points are brought in contact and then separated for a short distance, the space between them will be bridged over by an arc of blinding light. On examination, this light is found to be due to the intense whiteness of the carbon tips, chiefly, but *not to their combustion*, since in a vacuum, where combustion can not occur, the light is of equal intensity.

The heat of this arc of light is wonderfully intense. Platinum, more difficult to melt than other metals, melts in this heat like wax in the flame of a taper. Even quartz, and other bodies equally difficult to melt, are fused by it readily.

The electric light is produced not only by the "electric arc," but also by "incandescence." If a platinum wire, or a small rod of carbon, is placed in the circuit of a battery of high resistance, it will be heated intensely, and glow with a fine white light.

Electric lamps are of these two classes: some are constructed to yield the light by the arc, others to yield it by incandescence.

Magnetism. — Bars of soft iron inclosed in coils of wire are called ELECTRO-MAGNETS. The coil is generally called a HELIX. If the two ends of the coil be fastened to the poles of a battery, the electricity darts instantly through the coil, and *the iron becomes a magnet*. The bar of iron may be of any form; when in the shape of the horseshoe magnet, the coil is made in two parts, one encircling each arm of the iron. A horseshoe electro-magnet, A B, is seen in Fig. 153.

The strength of electro-magnets is something surprising. One belonging to Yale College, weighing 59 pounds, lifted a weight of 2,500 pounds. This wonderful power is developed only when an armature, c d (Fig. 153), is in contact with the poles. Without this, the magnet will not lift a tenth part of what it could otherwise sustain.

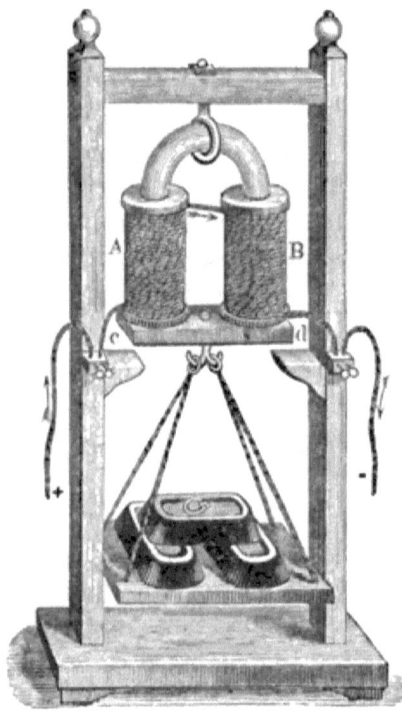

Fig. 153.

The iron is magnetic only while the electricity acts around it. Let the circuit be in any way broken, and the grasp of the giant is at once loosed; the load falls. On again making the circuit, the magnet is instantly as strong as before. The rapidity with which an iron bar will thus receive and part with its magnetism, as the circuit is made and broken, is truly astonishing. By the electric register for vibrations, the author has caused an electro-magnet to undergo this change at the rate of 8,400 times a minute.

The Electric Telegraph acts on this Principle. — It is upon this principle that the electric telegraph has enabled man to send his thoughts, with lightning-speed, across continents and under oceans, to his most distant fellowmen.

Having found that a bar of iron will become magnetic as

often as electricity is sent round it, and cease to be so on the instant the circuit is opened, let us next notice that the wires conveying the force may be of any length, even miles, and hence the battery may be in one city, while the magne, may be in another, and still an armature will be drawn against its poles every time the circuit is made. Now, if the motion of an armature to and from the poles can be made to *write*, then can messages be sent from one city to another.

The apparatus consists of three parts: the *key*, the *line*, and the *register*.

The Key is an instrument by which the circuit can be made and broken at will. It is in the office *from* which the message is to be sent. A brass lever, L (Fig. 154), moves on an axis, A. Two projections, n and m, from its lower side, are just above two others, one of which, a, is joined by a wire with the battery, while from the axis, A, another wire reaches to the distant station. By pressing the finger on the end of this lever, the point is brought in contact with the battery-wire at a, and the electricity can then act through the lever, from the battery-wire to the wire from the axis. Let the finger be lifted, and the lever will rise by the action of a spring, s, and the circuit is broken.

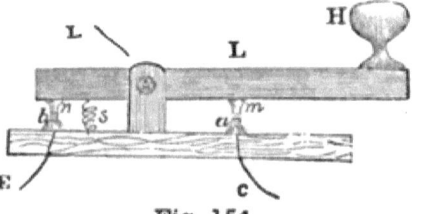

Fig. 154.

The Line consists of a wire, L, reaching from the key over the country to distant places. At first two wires were used, one from the positive pole, the other from the negative pole, of the battery; but it was soon found that the earth may take the place of one of these wires.

The Register is shown in Fig. 155. One of the screw-cups at the end of the instrument is connected with the line wire, L, which reaches from the key of the distant station, while the other, M, is connected with the earth. When the circuit is made, the electric force darts around the electro-magnet, and draws the armature down against its poles: this raises the long arm of the lever, and presses the steel point, I, against a strip of paper, which is pulled along from the spool, E, by clock-work. When the circuit is broken, the armature is released from the poles of the electro-magnet; the long arm of the lever falls by its own weight, or by the force of a spring, and the point is removed from the paper.

If the point press the paper for an instant only, a dot will

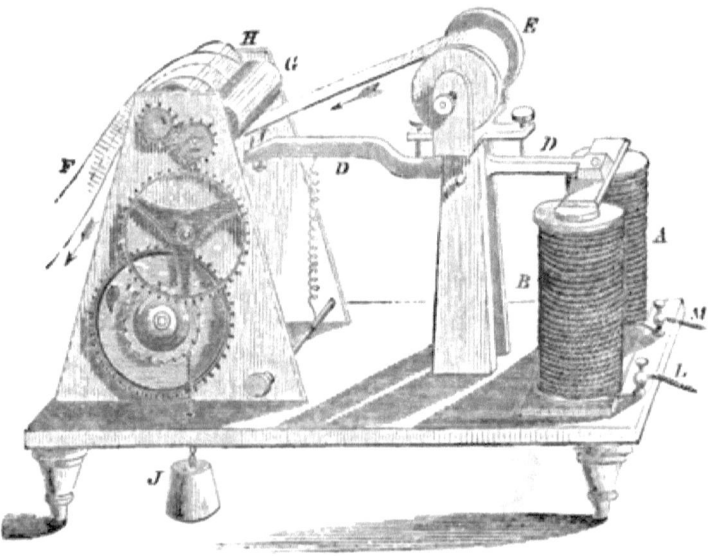

Fig. 155.

be made, but, if it be held against it for a longer time, a dash will be left upon it. Now, the letters of the alphabet are represented by dots and dashes. Two operators who know this alphabet can communicate with each other; one by pressing the key causes a series of dots and dashes to be marked upon the paper of the register at a distant place, while the other can read this written language. A skillful operator knows the letters by the *sound* of the clicks of the instrument. He uses a " sounder " instead of a writer.

Such is an outline of the *essential* parts of the electric

telegraph. A larger work, or, better, a visit to the telegraph-office, will make one fully acquainted with the many details of its operation.

The steam-engine and the electric telegraph may be regarded as the body and the spirit of modern civilization, the first distributing *matter*, the second *thought;* both laboring toward a more general diffusion of comfort and knowledge and sympathy among men.

Chemical Action. — Acidulated water will be decomposed by the current when the electrodes of a battery of two or more cells in series are immersed in it.

The electrodes of the battery (Fig. 156) pass up through the bottom of a vase into the very dilute acid, and two tubes

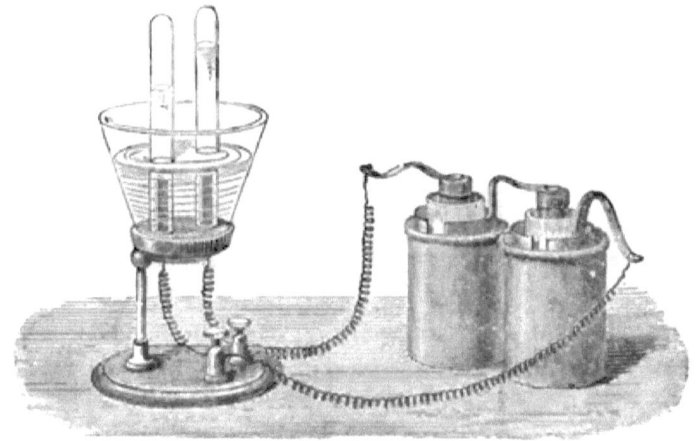

Fig. 156.

filled with the liquid are inverted over them. The moment the circuit is closed, a multitude of gas-bubbles break from the electrodes, and are caught in the tubes above. These gases prove to be hydrogen and oxygen, — the constituents of water.

Various other liquids may be decomposed by the current; all such are called ELECTROLYTES.

106. The current sways a magnetic needle near which it flows.

The galvanometer detects and measures the current on this principle.

Effect on the Needle. — Let a magnetic needle be placed in a rectangle of wire provided with pole-cups as shown in Fig. 157. It will be found that whenever the current flows through the wire, the needle will turn. *The current tends to put the needle at right angles to its own direction.*

Fig. 157.

If the current pass from south to north, *above* the needle, the north pole will turn to the *west;* if from north to south, the north pole will turn to the *east.* If the current pass *below* the needle, the north pole turns in directions opposite those just mentioned.

The *motion* of the needle declares the *presence* of a current; the *direction* of the motion tells the *direction* of the current; and the *distance* the pole moves measures the *strength* of the current.

The Astatic Needle. — This consists of two needles fastened together with the north pole of one opposite the south pole of the other. The ends of this needle are equally attracted and repelled by the earth's magnetism. But the north pole is usually left a little stronger than its companion, so that the earth's magnetism is not quite neutralized, and the needle will be "true to the pole," although with a very feeble force.

Fig. 158.

The Galvanometer. — A sensitive form of this instrument is shown in Fig. 158.

The astatic galvanometer consists of a coil of very fine silk-covered copper wire wound on a flat wooden bobbin; an astatic needle, the upper half near the upper plane-surface of the helix, the under one in the middle section of the helix; the needle is suspended by a silk fiber; a graduated circle is placed under the upper needle; the coil rests on a mahogany base with leveling screws, and is covered by a glass shade. (*Ritchie.*)

107. Induced or secondary currents are developed in a conductor: —

1st, By the approach or departure of a battery current.
This principle is embodied in the Ruhmkorff coil.
2d, By the approach or departure of a magnet.
This principle is embodied in magneto-electric machines and the telephone. (*G.* 874–876, 883, 886, 889.)

I. — INDUCTION BY A CURRENT.

Description of Apparatus. — Two coils, one containing a few feet of stout copper wire, and called the *primary* coil, and another containing a very great length of fine copper wire, and called the *secondary* coil, are so made that the primary may be placed inside the secondary or taken out at pleasure (Fig. 159). The primary is connected with a battery, and the secondary with a galvanometer.

Results of Experiment. — While the primary is *passing into* the secondary coil, the galvanometer needle swings, showing a current in the secondary moving in a direction *opposite to that of the battery current*. It is called an *inverse* current.

While the primary is being *withdrawn*, the needle swings the other way, showing a current in the secondary coil moving in *the same direction as the battery current*. It is a *direct* current.

But the current may be introduced into the secondary coil, and withdrawn in another way. We may leave the primary coil inside the secondary, and then simply close and open the battery circuit. On closing the circuit the current darts through the primary coil, and on opening it the current ceases. Now we find that on *closing the primary circuit*, the motion of the galvanometer declares an *inverse* current in the secondary coil. On *opening the primary* circuit, the motion of the galvanometer shows a *direct* current in the secondary coil.

Fig. 159.

Induction Coils. — It is upon this principle of induction that the *Ruhmkorff coils* are constructed. In these important instruments there is first a primary coil of large copper wire, inside of which is put a bundle of iron wires. Outside of this is placed the secondary coil, which is made of fine copper wire many thousands of feet in length. The two coils are insulated from each other with the utmost care. The ends of the primary coil are attached to the battery, while from the ends of the secondary coil the electricity is taken in experiments.

Ritchie's Induction Coil. — The induction coil, as constructed by Ritchie, is represented in Fig. 160. The primary consists of about 200 feet of stout copper wire. The secondary contains about 68,000 feet of fine copper wire. The two are separated by a thick glass bell whose knob is seen at the top. The primary circuit is opened and closed by a toothed wheel, *b*. The ends of the secondary are fastened to sliding rods on glass supports. When the poles of a battery are placed in the binding-posts *s* and *c*, let the circuit be opened and closed by turning the toothed wheel *b*, and electric sparks in rapid succession will leap through the air between the ends of the insulated rods. The length of the spark depends upon the size and insulation of the coils, and the strength of the battery. The instrument just described is able to yield a spark nine inches in length.

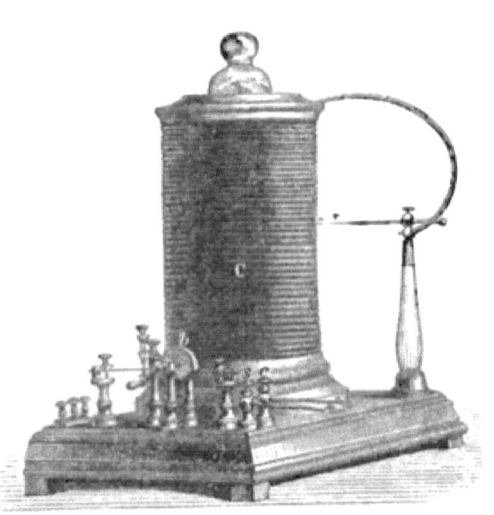

Fig. 160.

II. — INDUCTION BY A MAGNET.

Results of Experiment. — It is found that if a magnet be used instead of the primary coil in Fig. 159, the same effects are produced.

Hence the *motion of a magnetic pole induces a secondary current, in one direction as it enters, in the other as it emerges.*

If a bar of soft iron be placed in the secondary, and then a pole of a magnet be moved toward and from it (Fig. 161), the same effects will be produced. The soft iron becomes a magnet by the approach of the pole, and loses its magnetism when the pole departs. This is, in effect, only another way to introduce and withdraw the magnet.

Finally, insert a *permanent magnet* in the secondary coil, and move a bar or disk of *soft iron* toward and from its pole: currents in the coil are induced by this motion.

The soft iron becomes magnetic by induction, and then affects the magnet in the coil, *strengthening* it by approach and *weakening* it by departure.

Hence any *change in the strength of a magnet will induce an electric current in a coil which surrounds it.*

Fig. 161.

Magneto-Electricity. — Electricity induced by a magnet is called MAGNETO-ELECTRICITY. Many forms of apparatus have been devised with which to generate magneto-electricity, and the currents of enormous power, needed for electric lighting and other applications of electricity in the arts and industries, are produced by late and powerful forms of these "*dynamo-electric machines.*"

A Simple Form. — In this instrument (Fig. 162), two coils of wire, W, inclose the two arms of a bar of soft iron, having the form of a horseshoe magnet, which by a band and wheel, M, can be put in rapid motion in front of a very powerful compound magnet, S. The soft iron becomes magnetic whenever its ends are in front of the poles of the permanent magnet, and hence its two branches are being alternately magnetized in opposite states at every turn. The effect of

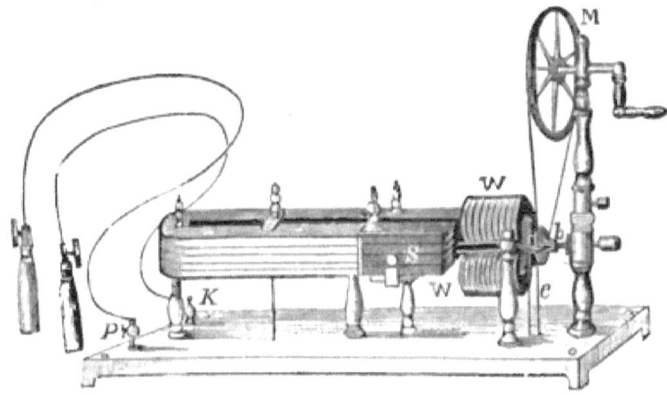

Fig. 162.

this is to produce two opposite currents in the coils at every revolution. These currents are taken by the wires, e, and thence under the instrument to the screw-cups, K and P.

The Telephone. — Any instrument for transmitting sound to a distance is a TELEPHONE. Two general classes of the instrument are in use, — *magneto* telephones, in which a magnet is used, and *electro-chemical* telephones, in which a battery is employed.

The Bell Telephone. — The Bell telephone belongs to the first class. Fig. 163 shows the inside structure, and Fig. 164 the external appearance, of the instrument.

B is a magnet, and A a coil of wire around one of its poles. C is a thin disk of sheet-iron, called the diaphragm, very near but not touching the pole of the magnet. F F are wires leading from the coil to the end of the handle, where one is joined to the line and the other to the earth. At the other end of the line is a second instrument of just the same kind.

Its Action. — A speaker puts his lips to the mouth of the cone of one instrument, and speaks, while a listener puts his ear to the mouth of the other, and hears what is spoken.

The air-waves of the voice vibrate the diaphragm C. Its motions toward and from the pole of the magnet strengthen and weaken the magnetism, and thus send electric pulses through the wire to the distant station.

The electric pulses from the *transmitting* telephone act through the coil of the *receiving* telephone at the other end of the line, and alternately strengthen and weaken its magnet. The diaphragm in front of this magnet is attracted by

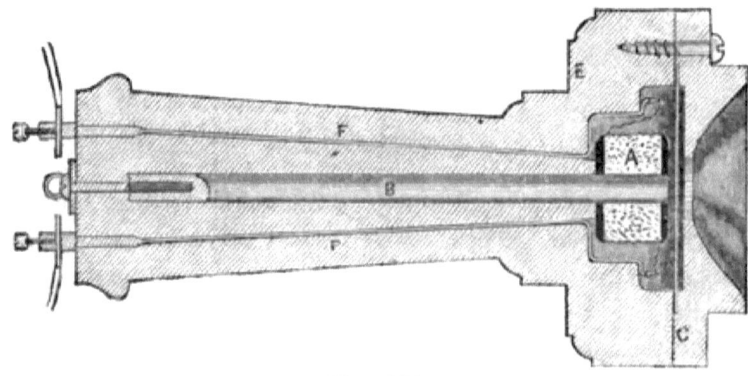

Fig. 163.

it, and, as the strength of the magnet varies, the thin disk springs back and forth. This vibration of the disk produces air-waves which enter the ear of the listener.

Now, two sets of air-waves which are exactly alike will affect the ear in exactly the same way, no matter how they are produced, and hence, all that is needed to make *any thing* speak is to cause it *to move so as to produce just such air-waves as the voice makes*. The air-waves of the voice of the speaker vibrate the iron plate in the *transmitter:* the iron plate in the *receiver* vibrates in exactly the same way, and hence the air-waves which enter the ear of the hearer are faithful copies of those which leave the lips of the speaker, and are heard as the same sound.

Hence, by the telephone, sound-waves are converted into electric pulses in the transmitter, and these electric pulses reaching the receiver are converted back again into sound-waves.

Edison's Carbon Transmitter. — In the Bell telephone we notice that the transmitter and receiver are exactly alike. But an Edison transmitter is often coupled with a Bell receiver. The telephone then becomes electro-chemical instead of magnetic.

The electric resistance of carbon varies with the pressure

Fig. 164.

upon it, doubtless, because heavier pressure secures a better contact of conductors. This is the principle used in the Edison transmitter.

Pure lampblack is formed into a compact button, and placed in a battery-circuit which also includes a Bell instrument at the distant station. A diaphragm is fixed so that its center presses gently against this button. When the diaphragm is at rest a *steady* current from the battery flows through the distant telephone. But, when the voice of the speaker vibrates the diaphragm, it exerts a varying pressure on the button, and this *varying pressure* transforms the steady current into an *undulatory* one. The diaphragm of the receiver is compelled to vibrate in unison with these undulations, and emit the sound.

The Microphone. — In the microphone the transmitter consists simply of a rod of carbon, A (Fig. 165), supported between two small blocks, C C, of the same material, its pointed ends resting loosely in shallow cups in the blocks. The whole is fixed to an upright thin board fastened to a solid base, D. This transmitter is put into circuit by joining one block to a battery by a wire Y, and the other by a wire X to the line leading to the distant receiver.

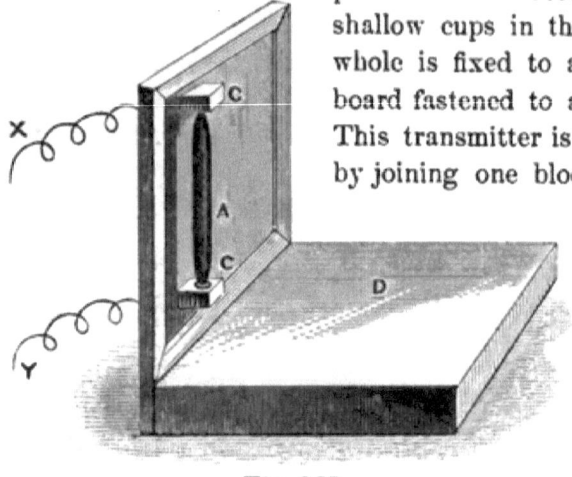

Fig. 165.

The ticking of a watch before this instrument is distinctly heard in the telephone, and even the steps of a fly over the board are heard by the distant listener.

108. Electricity may be developed by heat: it is then called THERMO-ELECTRICITY. (*G.* 900, 904.)

Thermo-Electricity. — We have seen that electricity will produce heat; we are now to notice that heat will produce electricity. When two pieces of different metals are soldered together, and their junction heated or cooled, a current of electricity is produced. The metals antimony and bismuth are best suited to this purpose, but any others, or, indeed, two pieces of the same metal, will, in some degree, produce the same effect. Nor is it quite necessary that metals should be used at all: other solids, and even fluids, give rise to this kind of electricity.

Two pieces of metal soldered together, with wires attached to their other ends, through which the electricity may act, is called a THERMO-ELECTRIC PAIR. When stronger currents are desired, a combination of pairs, the two metals alternating throughout, is used. Such a combination is called a THERMO-ELECTRIC PILE.

Metals which differ most in conducting power and crystalline texture, are best suited to produce thermo-electric currents. The force of the electricity is in proportion to the difference of temperature at the two ends of the pile, provided the difference does not exceed 80° or 90° F. It is in all cases very feeble; yet the galvanometer (Fig. 158) responds to so delicate a current that the slightest change of temperature in the pile can be detected by the electricity it produces. Indeed, the thermo-electric pile and galvanometer (Melloni's apparatus), is the most delicate thermometer known.

SECTION IV.

REVIEW.

I. — SUMMARY OF PRINCIPLES.

The electricity produced by friction shows its presence by attracting and repelling light bodies.

Two bodies charged with the same kind, either positive or negative, repel, but, when charged with different kinds, attract.

A charged body polarizes every other body in its neighborhood, inducing the same kind of electricity in all sides toward itself, and the opposite kind in all sides away from itself.

Electricity resides only upon the surface of a charged body.

Upon the surface of a sphere, electricity is uniformly distributed: if the body is not a sphere, the electricity will be most intense at the ends.

The *potential* of a body is the excess or defect of its

electric charge above or below that of the earth in its neighborhood.

The *electro-motive force* is that which urges electricity along over a conductor. The poorer the conductor, the greater the electro-motive force needed to carry electricity through it. In frictional electricity the electro-motive force is very great, as shown by its passage through air, which is among the very poorest conductors.

Two magnetic poles of the same name repel, but if of different names they attract, each other.

A pole of either name can not exist alone; but both are always found at once in the same piece of metal. The magnetic condition is always a *polarized* condition. There is no conduction of magnetism.

Current electricity is generated by chemical action in a galvanic cell, which may consist of any two different metals in a liquid which can act chemically on one more than on the other.

Current electricity differs from frictional electricity in its electro-motive force and quantity. Its electro-motive force is vastly less, its quantity vastly greater.

The + pole of a battery is always that one which is connected with the metal acted on *least* by the liquid.

Hydrogen is evolved by the chemical action, and clinging to the — plate "polarizes" it. The current on this account soon becomes weak.

A "constant battery" is one in which the hydrogen is prevented from reaching the — plate by means, usually, of a second liquid which changes the hydrogen into water.

The three most important elements of a current are its strength, C, its electro-motive force, E, and its resistance, R. Ohm's law is expressed in the formula $C = \dfrac{E}{R}$.

Each of these elements can be *measured* in terms of appropriate units.

The unit of R is called the OHM. A piece of No. 16

copper wire (about $\frac{1}{18}$ inch diameter) 60 feet long has a resistance of about one ohm. The resistance of ten times this length of the same wire is about ten ohms. (*G.* 908, 909.) The resistance of the Atlantic cable is eight thousand ohms.

The unit of E is called a VOLT. A single cell of Daniel's battery yields an electro-motive force of about one volt. A Grove or Bunsen cell is more powerful: its electro-motive force is about 1.8 volts.

The unit of C is called an AMPERE. A current of one Ampere strength will decompose .0000945 gramme of water in a second.

Electrical measurements are of the utmost importance in telegraphy and in other practical applications of electricity.

When great resistance is to be overcome, the cells of a battery should be joined " in series." When the resistance is small, the best effect is obtained by joining all plates of the same kind together.

The energy of the current may be converted into heat or light or magnetism or actinism.

Momentary currents are induced in a conductor by the making and breaking of a current in a conductor near it, or by the approach or departure of a magnet.

An undulatory current is induced in a conductor by the alternate strengthening and weakening of a primary current near it, or by the alternate strengthening and weakening of a magnet around which it is coiled.

II. — SUMMARY OF TOPICS.

89. Electricity by friction. — The electrical machine. — Its action. — Electroscopes. — Positive and negative. — The law. — Application of the law.

90. A charged body. — A non-conductor. — Potential. — High and low potential. — Electro-motive force. — Insulated body. — Polarization. — A charge by polarization.

91. Polarization in series. — Faraday's theory. — Differ-

ence between conductors and non-conductors. — Polarization before attraction. — Illustration.

92. Description of the Holtz machine. — Its action. — The explanation.

93. The Leyden-jar. — Various forms. — May be charged. — May be discharged. — Battery.

94. Electricity of the atmosphere. — Same as frictional electricity. — Lightning.

95. Effect of points. — Lightning-rods.

96. Mechanical effects of electricity. — Chemical. — Physiological.

97. Magnets. — Experiment. — Poles.

98. Attraction and repulsion.

99. Magnetic polarization. — In series. — Among the molecules.

100. A bar-magnet supported. — Points north and south. — Its variation. — Annual and daily changes.

101. Dip of the needle.

102. Different names for current electricity. — Chemical action. — With pure zinc. — The simple cell. — Chemical action in simple cell. — Electric conditions of the cell. — Direction of the current. — Enfeeblement of the current. — Remedied in Grove's cell. — What becomes of the hydrogen? — The Bunsen cell. — Other forms.

103. Statement of Ohm's law. — Definitions. — The formula. — Units of C, E, and R (see Review Summary).

104. Battery of low resistance. — Its electro-motive force. — Strength of current. — Battery of high resistance. — Its electro-motive force. — The resistance. — Strength of current. — Quantity and intensity.

105. Electricity produces heat. — And light. — And magnetism. — The electric telegraph. — The key. — The line. — The register. — Electricity produces chemical action.

106. The current deflects a magnetic needle. — The astatic needle. — The galvanometer.

107. Description of apparatus for induction by a current.

— Results of experiment. — Induction coils. — Ritchie's. — Experiment with a magnet. — Magneto electricity. — Simple form of machine. — The telephone. — The Bell telephone. — Its action. — Edison's carbon transmitter. — The microphone.

108. Thermo-electricity.

CHAPTER IX.

ON MACHINERY.

SECTION I.

ON THE SIMPLE MACHINES.

109. THE principle of work, applied to any one of the simple machines, will determine its law of equilibrium.

The Principle of Work. — We remember that *work* is the overcoming of resistance, and that it is measured by the product of the weight by the vertical height through which it is lifted. The *principle of work* states briefly that *two forces, acting in opposite directions upon the same body, will be in equilibrium when they do equal amounts of work*.

Let us illustrate a single case by means of Fig. 166. Suppose two bodies, M and N, hang from the ends of a bar, A B, which rests upon the point C, about which it may freely turn. If it does turn, and M goes up, N will go down, and if the distance B C is twice the distance A C, then N will go twice as far as M. In all cases, the vertical heights through which the two bodies will move have the *same ratio* as their distances, A C and B C, from the center of motion. *These lines may then be used* instead of the vertical heights in calculating work. Then the work which may be done by M will be M × A C, and that of N will be N × B C. Now, if these works are equal, then the two bodies will be exert-

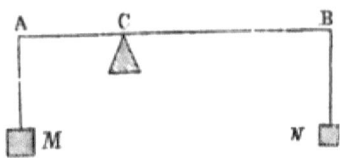

Fig. 166.

ing equal forces upon the bar A B, and, if once brought to rest, they will just balance each other.

Machines. — Machines are instruments by which forces may be applied to overcome resistance, or do work. They are so made that a small force, by moving rapidly, may overcome a greater resistance, or a great force, by moving slowly, may put a small resistance in rapid motion. In all cases the work done by the two forces must be equal.

The resistance to be overcome is always called the WEIGHT: the force which overcomes it is called the POWER.

Simple Machines. — There are six simple forms of machines, usually called the Mechanical Powers. Out of these six simple machines all forms of machinery, complex as they may be, are made. We name them in the order which is to be followed in describing them.

1. The Lever.
2. The Wheel and Axle.
3. The Pulley.
4. The Inclined Plane.
5. The Wedge.
6. The Screw.

The Law of Equilibrium. — By the term, law of equilibrium, is meant a statement of the relation which must exist between the power and the weight, in order that, when at rest, they may just balance each other.

110. Levers are of three classes. The principle of momentum, applied to the lever, shows that the power and weight will be in equilibrium when they are to each other inversely as the perpendicular distances from the fulcrum to the directions in which they act. A compound lever acts on the same principle. Applications of the lever are very numerous.

Levers. — A lever is an inflexible bar, able to turn freely upon one point.

Fig. 167.

Thus, if the line A B (Fig. 167) represents an inflexible bar, resting upon some support at F, upon which it has free motion, it represents a lever. The point F, about which the lever turns, is called the FULCRUM.

Three Classes of Levers.—That point of a lever to which the power is applied is called the POINT OF APPLICATION. That on which the weight acts is called the WORKING POINT. Now, the lever takes different names according to the relative positions of the point of application, the working point, and the fulcrum.

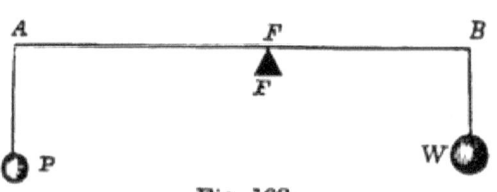

Fig. 168.

In the lever represented in Fig. 168, whose fulcrum is at F, a power, P, acts upon one end of the lever, A, while a weight, W, acts upon the other, B. The fulcrum is between the point of application and the working point. This is called a lever of the *first class*.

In the lever (Fig. 169), the working point, B, is between the point of application, A, and the fulcrum, E. This is a lever of the *second class*.

In the lever (Fig. 170), the point of application, A, is between the working point, B, and the fulcrum, F. This is a lever of the *third class*.

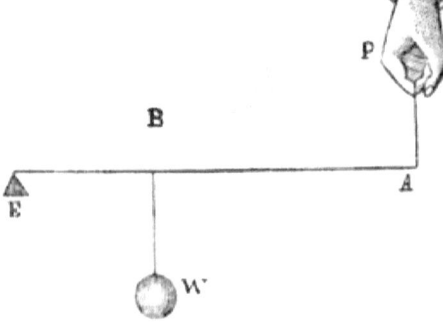

Fig. 169.

All levers belong to these three classes. They need not, however, be made in the simple straight form shown by the figures. In Fig. 171. the line A F B represents a lever whose arms make a right angle at the fulcrum, F. It is a lever of the first class; so also is that shown in Fig. 172 by the curved line A F B.

Application of the Principle of Work.—Now, if we

examine the figures which represent the three classes of lever, we see that in each one the power, P, and the weight, W, are two forces which act in opposite directions. They will be able to just balance each other, when of such strength that, if in motion, they would do equal amounts of work.

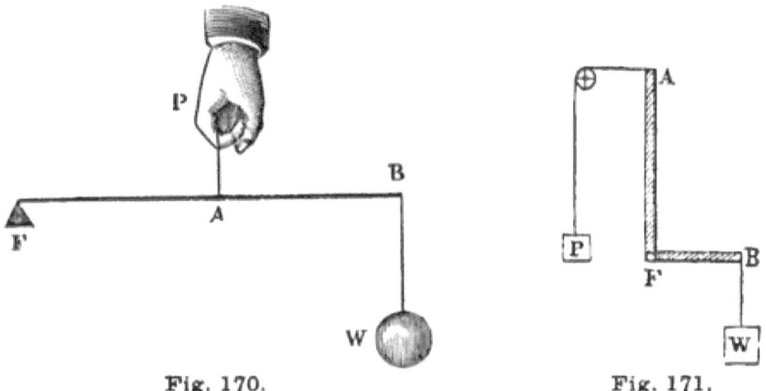

Fig. 170. Fig. 171.

The lines B F and A F have the *same ratio* as the *vertical heights* through which they move if motion occurs. The work of the power is therefore P × A F; that of the weight is W × B F. If equilibrium takes place only when the amounts of work are equal, then

P × A F = W × B F; hence,
P : W :: B F : A F.

This proportion teaches us that *the power and weight will be in equilibrium, when they are to each other inversely as the distance of their points of application from the fulcrum.*

It may be, however, that the power and weight do not act perpendicularly upon the lever. This case is represented by Fig. 172. The lever A B has its fulcrum at F. The power, P, and the weight, W, act obliquely at B and A. Now, it is evident that the force of the power, acting obliquely at B, is not all expended to lower the lever, but that if it were acting upon the point N, perpendicularly, it would exert all its forces to move the arm N F. So the effect of the weight acting obliquely upon A will be the same as if it were acting perpendicularly upon an arm, M F. Hence P × N F may be taken as the work of the power, and W × M F as the work of the weight. Putting these equal,

P × N F = W × M F; hence,
P : W :: M F : N F.

Law of Equilibrium. — This proportion teaches that

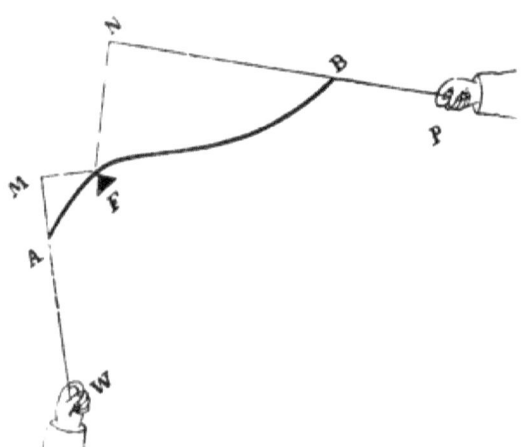

Fig. 172.

the power and weight will be in equilibrium, when the power and weight are inversely as the perpendicular distances from the fulcrum to the directions in which they act.

This principle is called the LAW OF EQUILIBRIUM for the lever. It will apply to all possible forms.

The Compound Lever. — In a compound lever several simple levers are generally so fixed, that the short arm of one may act upon the long arm of another. Fig. 173 shows a compound lever made up of two simple levers having their fulcrums at F and F'.

In this case the work of the power will be equal to P × C F × B F', and that of the weight will be equal to W × D F' × A F. If these products are put

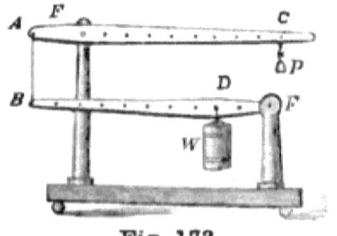

Fig. 173.

into the form of an equation it will be seen that *the power and weight will be in equilibrium, when the power multiplied by the product of all the arms on its side of the fulcrum is equal to the weight multiplied by the product of all the arms on its side.*

The compound lever is used when it is not convenient to have a very long lever, and yet a small force is required to sustain a very large weight. If the long arms of the two simple levers be six and eight feet, and each short arm is one foot, then one pound power at C will balance 48 pounds at D; while if a simple lever had been used whose long arm was as long as those two long ones together, $6+8=14$ feet, and whose short arm was one foot, then one pound at C would only be enough to balance 14 pounds at D.

Applications of the Lever. — Of levers of the first kind many familiar examples might be named. The hand-spike and crow-bar are levers of this class. Shears and pincers are pairs of levers, also of the first class; their fulcrums being at their joints.

The balance is one of the most useful applications of the lever. Fig. 174 represents a

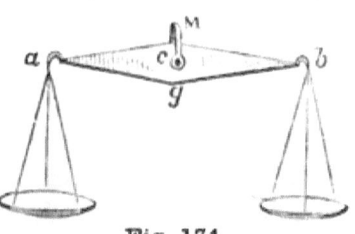

Fig. 174.

very simple form of this instrument. The beam $a\,b$ is a lever poised at its center, the pivot or fulcrum c being a *little above* its center of gravity. From the ends of the beam the scale-pans are hung, in one of which is put the body to be weighed, and, in the other, the weights to balance it. Balances are of continual use in commerce; they are indispensable in the laboratory of the chemist, for whose use they are made with so great skill that a weight equal to the $\frac{1}{10000}$ of a grain can be easily weighed.

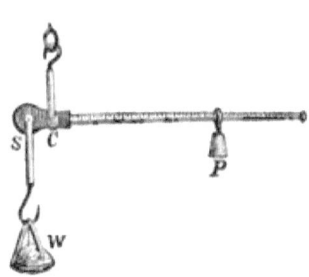

Fig. 175.

The steelyard is also a lever of the first class, but with unequal arms. The body W, Fig. 175, to be weighed, is hung from the short arm of the lever S B, and it is balanced by a small weight, P. It is clear that this small weight will balance more weight in the body W, as it is moved farther and farther from the fulcrum C. The arm B C has notches cut upon it, and numbered, to denote the pounds or ounces in W, balanced by P, when at these points.

Levers of the second class are not so common; the *wheelbarrow*, however, is an example sufficiently familiar. The axle of the wheel is the fulcrum; to the opposite ends of the handles the power is applied, while the load, or the weight, rests between these points. The oar of a boat is a lever of this kind, where, singularly enough, the unstable water serves as a fulcrum; the hand is the power at the other end of the lever, while the boat is the weight between them.

Levers of the third class are often met with in the arts. The common fire-tongs and the sheep-shears are pairs of levers of this kind. Their fulcrums are at one end; the resistance to be overcome is put between their parts near the other end, while the fingers, which afford the power, are between the fulcrum and the weight.

111. The wheel and axle acts on the principle of a lever. The power and weight will be in equilibrium when the power is to the weight as the radius of the axle is to the radius of the wheel.

A compound wheel and axle acts on the same principle as a compound lever.

One wheel may be made to turn another by friction, by cogs, or by bands.

Applications of this machine are common and important.

The Wheel and Axle. — One form of the wheel and axle is shown in Fig. 176. It consists of a wheel, B, firmly fastened to an axle, A, and turning freely around an axis, one end of which is shown at C. The power, P, acts upon the circumference of the wheel, and the weight, W, acts upon the axle by means of a rope winding around it in the opposite direction.

It acts on the Principle of the Lever.—If we have an end view of the machine, it will be seen, as shown in Fig. 177, where the large circle represents the wheel, and the small circle the axle; the center, C, being the end of the axis. At the point A, the power acts on the wheel, and from the point B, on the other side of the center, the weight is suspended. Now, if a straight line, A B, join the

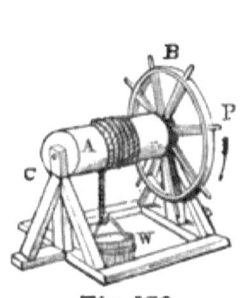
Fig. 176.

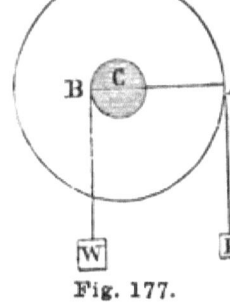
Fig. 177.

points A and B, it will pass through the center, and represent a lever, whose fulcrum is at C. It is upon the ends of such a lever that the power and weight are constantly acting.

Application of the Principle of Work.—The work of the power is represented by $P \times A C$; that of the weight by $W \times B C$. If the two forces are able to balance each other, these products are equal. Hence,

$$P \times A C = W \times B C; \text{ or,}$$
$$P : W :: B C : A C.$$

But, in the figure, we notice that A C is the radius of the wheel, and that B C is the radius of the axle. Then the proportion teaches us that *the power and weight will be in equilibrium when the power is to the weight as the radius of the axle is to the radius of the wheel.*

If the radius of the axle is one foot, and that of the wheel three feet, then one pound upon the wheel will balance three pounds upon the axle.

The Compound Wheel and Axle. — When more than one wheel and axle are connected, so that the axle of each may act on the wheel of the next, the machine is a compound wheel and axle. Such an arrangement is shown in Fig. 178. We may get the law of equilibrium in the same way as in the compound lever. The work of the power will be P, multiplied by the several radii of the wheels; that of the weight will be W, multiplied by the several radii of the axles. If the two forces are able to balance each other, these values must be equal. Hence we learn that in a compound wheel and axle, the power and weight will be in equilibrium, when *the power multiplied by the product of the radii of the wheels equals the weight multiplied by the product of the radii of the axles.*

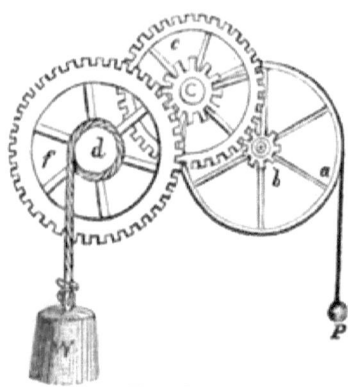

Fig. 178.

It is easy to see that in this way a small power may be made to balance a much larger weight than it could by acting upon a simple wheel and axle, unless the wheel should be so large as to be unwieldy.

One Wheel may turn another by means of Cogs. — In Fig. 178, there may be seen projecting teeth on the circumferences of the axles, *b* and *c*, which fit into equal notches on the circumferences of the wheels. Neither the axles nor the wheels can turn without causing the other to turn also. This is the common and convenient method of giving motion from one wheel to another. The wheels of a clock are cog-wheels: those of a watch also beautifully illustrate this mode of communicating motion.

By Friction. — When the circumferences of the wheels and the axles are made smooth, they may be pressed so snugly together, that neither can turn without turning the other at the same time, in the opposite direction. In this case, the motion is communicated by the *friction* of the parts against each other.

By Bands. — A third method of giving motion to a train of wheel-work consists in the use of bands or belts, which encircle the parts which are to act upon each other. In the spinning-wheel, for example, the spindle is turned by a band which passes around it and the axle of the wheel-head. Another band passes around the wheel-head and the large wheel, which is turned by the hand of the spinner. From the horse-power of a threshing-machine, also, motion is given to the cylinder by means of a belt.

Applications of the Wheel and Axle. — Many forms of the wheel and axle are in common use: the windlass is one of the most familiar, being often used to raise water from wells. One form of the windlass is represented in Fig. 176. A crank is often used in place of the wheel, B. The common grindstone is a homely illustration of the wheel and axle: the crank is in place of a wheel; the stone itself is the axle. The power is the force of the hand, while the weight is the resistance offered by the tool pressing on the edge of the stone.

If the axle is in a vertical position, and the forces of power and weight act horizontally, the machine is then called a CAP-STAN, and is much used on board of ships.

The compound wheel and axle is used in almost every mill and factory. Two objects are sought in its use : either great resistance is to be overcome, or rapid motion is to be secured. To overcome great resistances, the power is applied to the circumference of the first wheel in the system, and the weight is acted upon by the last axle. This case is shown in Fig. 178. To secure rapid motion, the power is applied to the first axle, while the weight is acted upon by the circumference of the last wheel. The same figure illustrates this case also, if we will suppose the heavy body, W, to act as a power to put the lighter body, P, in motion. If we suppose the radius of each axle to be one foot, and of each wheel ten feet, then

$P \times 10 \times 10 \times 10 = W \times 1 \times 1 \times 1$; or, $1,000\ P = W$. Now, W being 1,000 times heavier than P, P must move 1,000 times faster than W. In this way, a great power may be changed into rapid motion. An example of this is found in the saw-mill, where the slow motion of a heavy body of water, acting against a water-wheel, is given, by means of cogs and belts, from wheel to wheel, until it re-appears, multiplied a thousand-fold, in the buzzing saw.

112. The pulley may be either fixed or movable. In the fixed pulley the power and weight will be in equilibrium when they are equal.

In the movable pulley, with a single rope, the power and weight will be in equilibrium when the power is equal to the weight, divided by the number of branches of rope which sustains the weight.

In movable pulleys, with separate ropes, the power and weight will be in equilibrium when the power equals the weight, divided by two raised to a power shown by the number of pulleys.

The applications of the pulley are common and important.

The Pulley.—A pulley is a grooved wheel, turning freely about its axis, with a rope passing over or around it. It is shown in Fig. 179. The grooved wheel, A, moves freely upon its axis, while over its circumference goes the rope, to the ends of which the power and the weight are fastened.

Is either Fixed or Movable.—If the axis of the pulley is stationary (see Fig. 179), the pulley is called a FIXED PULLEY. One whose axis moves with the weight is called a MOVABLE PULLEY. This will be understood by means of Fig. 180. The wheel, E, is a movable pulley. From its axis the weight is hung, while the rope, one end of which is fastened to a fixed support at D, passes under it, and then over a fixed pulley, A. The power is applied to this end of the rope.

The Principle of Work applied to the Fixed Pulley.
— The fixed pulley is shown in Fig. 179, to which we again refer. It is clear that, when motion occurs, the power, P, will go down exactly as far as the weight, W, goes up. To do equal work when the distances are equal, the bodies must have equal weights. Hence, *in the fixed pulley the power and weight can balance each other only when they are equal.*

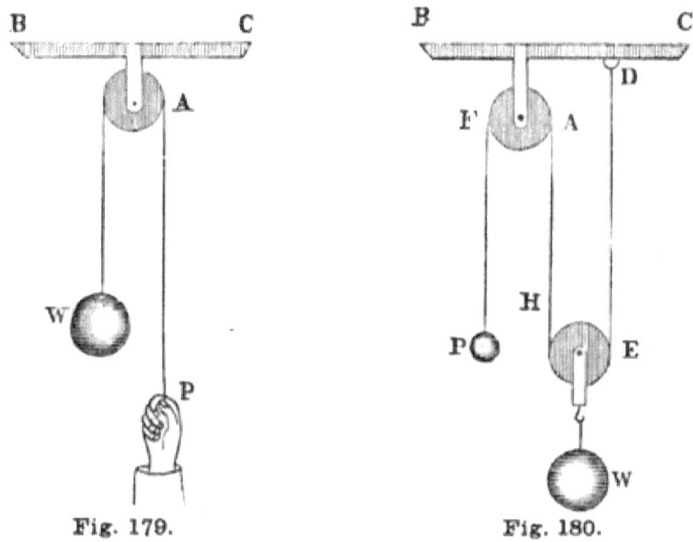

Fig. 179. Fig. 180.

The Principle of Work applied to the Movable Pulley with a single Rope. — In the movable pulley with a single rope (see Fig. 180), the weight rests upon two branches of the rope, H and E; and, when it rises, both branches must be equally shortened. But the rope, F P, will lengthen just as much as both the branches shorten. It is clear that the power, P, must move down just twice as far as the weight, W, goes up. Let V represent the vertical height through which the power moves, then $\frac{V}{2}$ will represent the vertical height through which the weight is lifted. The work of P will be represented by $P \times V$, and that of W by $W \times \frac{V}{2}$, and the two forces will balance each other when
$$P \times V = W \times \frac{V}{2}, \text{ or when } P = \frac{W}{2}.$$

Now let us take another case. Suppose there are two movable pulleys, C and D (Fig. 181), with a single rope, one end being fastened at F, while to the other end the power, P, is applied. In this case we find that the weight is supported by *four* branches of the rope, and we see, too, that, when it rises, all four of these branches must be shortened alike. But the rope, E P, must at the same time lengthen as much as all the branches shorten, so that the distance of P downward must be four times as great as that of W upward. Then, if V is the vertical height for P, $\frac{V}{4}$ will be the vertical height for W; and, if they do equal work,

$$P \times V = W \times \tfrac{V}{4}, \text{ or } P = \tfrac{W}{4}.$$

In like manner, if three movable pulleys are used, we shall find that, to be in equilibrium, $P = \tfrac{W}{6}$.

Fig. 181.

If, now, we notice that in each of the values of P just found, the denominator of the fraction is the number of branches of the rope which supports the weight, we have this general principle: in movable pulleys, with a single rope, *the power and weight will be in equilibrium when the power equals the weight divided by the number of branches which support it.*

The Movable Pulley with separate Ropes. — When each pulley has a separate rope, the law is very different. Fig. 182 shows such a system. The three ropes, $d\,f\,h$, are fastened to the beam. The first, after passing around the pulley, $b\,d$, is fastened to the axis of the one above: so the rope f, after going around the pulley $e\,f$, is fastened to the axis of $g\,h$; but the rope h, after going under the pulley $g\,h$, passes over a fixed pulley, and receives the power at the other end.

The Principle of Work applied to the Movable Pulley with separate Ropes. — This system is only a *combination of movable pulleys with a single rope*. Suppose the pulleys bd and ef were taken away, the weight being hung from the axis of gh. There would be left an arrangement exactly like that shown in Fig. 180. gh is a movable pulley, with a single rope, to lift the pulley ef, which is likewise a movable pulley, with a single rope, to lift the pulley bd; while bd is itself a movable pulley, with a single rope, to lift the weight, W. The effect of the power, P, will be doubled by each pulley thus: —

With 1 pulley, $P = \frac{W}{2} = \frac{W}{2^1}$;
" 2 pulleys, $P = \frac{W}{4} = \frac{W}{2^2}$;
" 3 pulleys, $P = \frac{W}{8} = \frac{W}{2^3}$.

If we notice that the denominator, in each of these values of P, is a power of two, whose *index* is the *number of pulleys*, we infer that, in a system of movable pulleys with separate

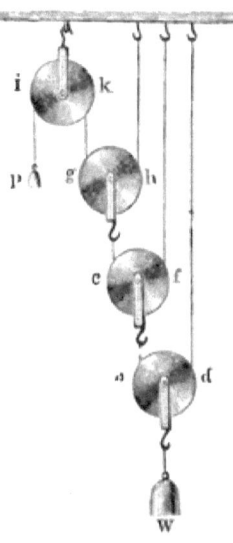

Fig. 182.

ropes, *the power and weight will be in equilibrium when the power equals the weight divided by a power of two, whose index is the number of pulleys.*

For example, with a system of five pulleys, how much weight will a power of ten pounds balance?

$P = \frac{W}{2^5}$; or $10 = \frac{W}{32}$; hence $W = 320$ pounds.

Applications of the Pulley. — No mechanical advantage is gained by the use of the *fixed pulley*, because the weight must move just as fast as the power, yet it is of great value in the arts, for changing the direction of forces. A sailor standing upon the deck of his ship may, by the use of a fixed pulley, hoist the sail to the top of the loftiest mast; or when heavy bales or boxes are to be lifted to the upper floors of warehouses, a horse, trotting along the level yard or street (Fig. 183), will lift them as effectually as though he were able to climb the perpendicular wall with the same rapidity.

The Movable Pulleys with single rope are in common use for moving heavy weights through considerable distances. Merchandise may be lifted by means of them, from the hold of a ship to the wharf, or to the upper stories of storehouses; or, by a different arrangement of the machine, the ship itself may be drawn from the water for repairs. In practice, the fixed pulleys of a system are placed side by side, and thus form what is called a BLOCK: the movable pulleys, likewise side by side, form another block. By this means the system is made compact.

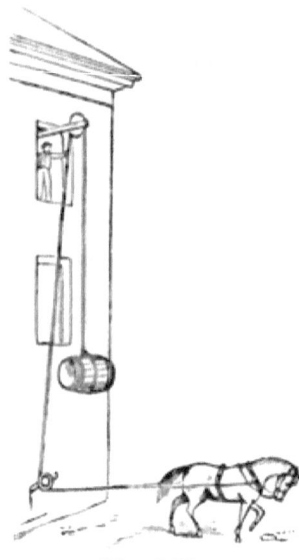

Fig. 183.

In all pulleys there is a loss of power, due to the friction of the pulleys in the blocks, to the weight of the lower block, and to the stiffness of the ropes used; so that the weight, actually overcome by a given power, is always less than the laws of equilibrium would afford.

113. The principle of work applied to the inclined plane shows: —

1st, That, when the power acts parallel to the length of the plane, the power and weight will be in equilibrium when the power is to the weight as the height of the plane is to its length;

2d, That, when the power acts parallel to the base of the plane, the power and weight will be in equilibrium when the power is to the weight as the height of the plane is to its base.

The applications of this machine are very numerous.

The Inclined Plane. — Any plane, hard surface, placed in an oblique position, may be used as an inclined plane. In

Fig. 184, A B represents an inclined plane. The distance B C is the *height* of the plane, and A C is its base. The weight, W, may be urged up the plane by a force acting parallel to A B, or parallel to A C, or at any angle to these. We are to notice the first two cases only.

If the Power acts parallel to the Length of the Plane. — In the figure the power, P, by means of a rope going over the fixed pulley, D, at the top of the plane, acts upon the weight, W, in a direction W D, parallel to the length, A B, of the plane.

Now, a force which will urge the weight from A to B is *lifting* it only through the vertical height, C B.

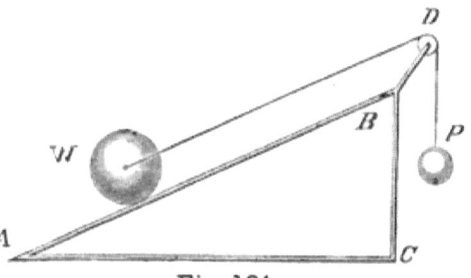

Fig. 184.

But, while the weight goes from A to B, the rope passing over the pulley will let the power down a distance equal to A B, in the *same time*. The work of the power is, therefore, represented by P × A B; that of the weight, by W × C B. When these products are equal, the two forces will be able to balance each other. Thus: —

P × A B = W × C B; or
P : W :: C B : A B.

This proportion teaches us that, *when in equilibrium, the power is to the weight as the height of the plane is to its length.*

If, for example, the height C B is four feet, and the length of the plane, A B, is sixteen feet, a power of one pound will balance a weight of four pounds. For —

1 lb. : 4 lbs. :: 4 ft. : 16 ft.

If the Power acts parallel to the Base of the Plane. — Let A B (Fig. 185) represent a plane whose height is C B, and whose base is A C. The power acts upon the weight by means of a cord passing over the pulley at C, in a direction parallel to A C. To move the weight from A to B, will be lifting it only through the *vertical height, C B.* If the pulley, C, could be raised while the weight goes up, so as to keep the cord parallel to A C, then the cord, passing over the pulley, will let the power down a distance equal to A C. The work of the power is, therefore, represented by P × A C, and that of the weight, by W × C B. If now these products are equal, the power and weight will just balance each other. Hence —

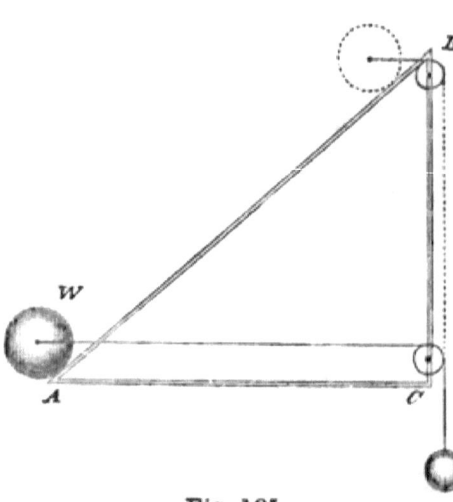

Fig. 185.

P × A C = W × B C; or
P : W :: B C : A C.

From this proportion we learn that, when the power acts parallel to the base of the plane, *the power and weight will be in equilibrium when the power is to the weight as the height of the plane is to its base.*

Thus, if the height of the plane is two feet, and the base is ten feet, a power of one pound will balance a weight of five pounds. For 1 lb. : 5 lbs. :: 2 ft. : 10 ft.

Applications of the Inclined Plane. — This machine is used to lift heavy weights through short distances. Many familiar examples might be named. If a barrel of merchandise is to be placed upon a wagon, it is often rolled up on a plank, one end of which rests on the ground, the other on the wagon. A hogshead which a dozen men could not lift may thus be raised by the strength of one or two.

Our common stairs are, in principle, inclined planes, the

arrangement of steps only giving a firm footing. If the distance between the floors be three-fourths the *length* of the stairs, then, besides the ordinary effort of walking, the person must continually, while going up, labor to lift three-fourths of the weight of his body.

114. The wedge, in its most common form, is made up of two inclined planes joined together at their bases. The sharper the wedge, the greater the resistance which may be overcome by it.

The Wedge. — This instrument is shown in use by Fig. 186. A B is called the back of the wedge: A c and B c are its sides, and c is its edge. It is generally used in cleaving timber, and sometimes for raising heavy weights through very short distances. For these purposes its edge is put into a crevice made for it, and it is then driven by blows with a sledge.

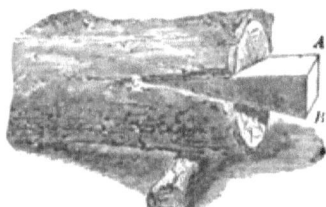

Fig. 186.

Since we can not calculate the force of a blow, no attempt is here made to establish a law of equilibrium for the wedge.

115. This principle applied to the screw shows that: —

The power and weight will be in equilibrium when the power is to the weight as the distance between two contiguous threads is to the circumference of the circle in which the power moves.

The screw is used extensively to produce great pressure. It is also often used to measure delicate distances.

The Screw. — The screw consists of a cylinder of wood or metal, with a spiral groove winding around its circumference. This grooved cylinder (C, Fig. 187) passes through a block, N G, on the inside surface of which is a spiral groove, into which the raised parts of the cylinder exactly fit. The block is usually called the *nut*. The raised part between the grooves of the cylinder is called the *thread*.

Suppose the nut to be stationary: then, if the screw is turned by a power acting upon the lever at B, it must advance downward at every revolution, and the pressure of the advancing screw will be exerted upon any object placed under the press-board, E F, against which the end of the screw presses.

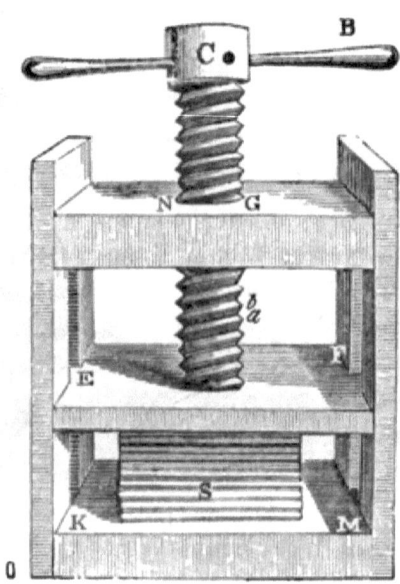
Fig. 187.

Application of the Principle of Work. — By one turn of the screw, it will advance downward a distance just equal to the distance between two contiguous threads. The press-board, E F, which may be regarded as the weight, will be moved along through an equal distance, $a\ b$, by every turn. The power acting at B will, in the same time, move through the circumference of the circle whose radius is B C. Hence the work of the power will be represented by P × circumference of the circle whose radius is B C, and that of the weight by W × $a\ b$. If these products are equal, the two forces, when at rest, will be in equilibrium. Hence: —

P × circ. B C = W × $a\ b$; or
P : W :: $a\ b$: circ. B C.

This proportion teaches that *the power and weight will be in equilibrium when the power is to the weight as the distance between two contiguous threads is to the circumference of the circle in which the power moves.*

Thus, if the distance between the threads is $\frac{1}{2}$ inch, and the circumference traveled by the power is 5 feet, or 60 inches, what weight on the nut would one pound power at B balance?

1 lb. : W :: $\frac{1}{2}$ in. : 60 in. W = 120 lbs.

Applications of the Screw. — The screw is used when great weights are to be lifted short distances, or when heavy pressure is to be exerted. By its use, cotton is pressed into bales, the juices of fruits extracted, and oils pressed from vegetable bodies such as linseed and the almond.

Micrometers. — In contrast with these uses of the screw, depending on the immense pressure it can exert, is another, remarkable for its delicacy. It is used to measure very small distances when accuracy is required. Screws with threads of exceeding fineness, and called MICROMETERS, are used for this purpose. Suppose a screw with a hundred threads in one inch of its length; then, at every turn its end would advance just $\frac{1}{100}$ of an inch, and, if it carry a steel marker, spaces of that length may be marked off on any body alongside of which it moves. Now, let the power move in a circle ten inches in circumference, and let this circle be graduated to inches, tenths, and hundredths. If the power move one inch on this scale, the marker on the end of the screw will go forward only $\frac{1}{1000}$ of an inch. If the power goes $\frac{1}{10}$ inch, then the marker will advance only $\frac{1}{10000}$ of an inch, a distance quite too small to be seen except by the aid of a good microscope. Astronomers use the micrometer screw, to measure the apparent sizes of the heavenly bodies.

SECTION II.

ON WATER-POWER.

116. Water-wheels are turned by the power of moving water. There are several kinds: first, the Undershot wheel; second, the Overshot wheel; third, the Breast wheel; fourth, the Turbine wheel.

The Undershot Wheel. — The undershot wheel is shown in Fig. 188. Its circumference is provided with *float-*

boards, a d c, against which the running water acts. Other wheels are connected with the axle of this one by cogs and bands. This form of wheel is often placed in a horizontal position, and water from the bottom of a dam guided against the float-boards of one side.

Fig. 188.

The Overshot Wheel. — The overshot wheel (Fig. 189) differs from the undershot, by having buckets upon its circumference, instead of float-boards. The water enters the buckets at the top of the wheel, and, filling those on one side of it, turns the wheel by its weight. The buckets all open in the same direction, so that while those on one side of the wheel are full, those on the other side will be bottom upward and empty.

The Breast Wheel. — The breast wheel (Fig. 190) differs from the undershot wheel only by being so placed in front of a dam, that the water shall fall upon the float-boards of its circumference on a level with its axis.

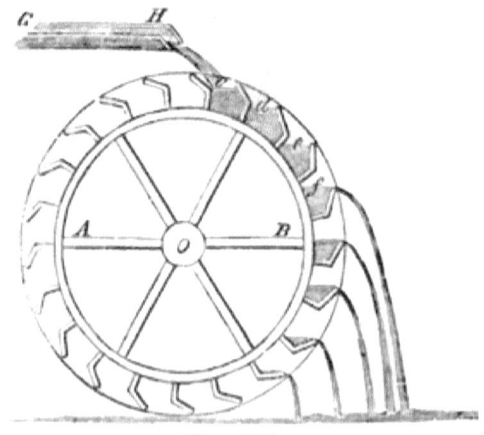

Fig. 189.

The American Turbine. — The construction of the turbine is more complex than that of the wheels just described. Its action may be understood by a careful study of Fig. 191.

The figure shows a *section* of the *interior* of the wheel, as it would appear to one who looks down upon it as it lies in its horizontal position. In the center is a circular disk of cast iron, A B, in a horizontal position. On the upper surface of this disk are fastened the curved guides, *a a a*. *This disk is stationary.* The wheel proper, C D, revolves outside of this disk. It consists of two cast-iron plates, one above the other, the space between them being divided into numerous channels by the curved partitions, *c c c*. The partitions in the wheel, and the guides on the disk, are curved in opposite directions. To the bottom of this wheel is fastened a cast-iron plate, which extends *under* the central disk, A B, and to the center of this plate is attached a vertical shaft which comes up through the disk at E. The revolving part, therefore, consists of the outside wheel, D, the iron plate underneath, and the vertical shaft, E.

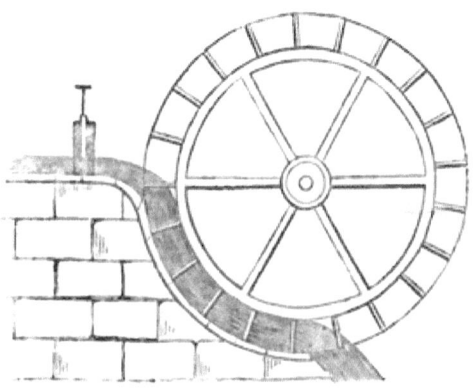

Fig. 190.

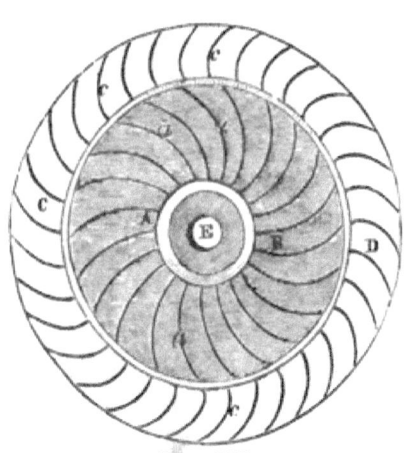

Fig. 191.

The turbine is placed at the bottom of a column of water. The weight of the water above the disk forces the water with great power out from between the curved guides, *a a*, into the curved channels, *c c*, of the wheel. *The energy of these streams* turns the wheel in the direction in which they strike against its partitions. The vertical shaft turns with the wheel, and, by means of cogs, gives motion to other parts of the machinery.

Of all forms of water-wheel, the turbine is most energetic and economical.

SECTION III.

ON THE STEAM-ENGINE.

117. The elastic force of steam is applied to mechanical purposes by means of a steam-engine. The essential parts of this machine are, 1st, the boiler in which steam is generated; 2d, the cylinder in which the steam is made to move a piston; 3d, the crank by which the piston turns a wheel. Engines are either high-pressure or low-pressure.

The Tension of Steam. — When steam is formed at a temperature of 212°, its elastic force or tension is just equal to the pressure of the atmosphere, or 15 pounds to the square inch. If taken out into another vessel, preserving its temperature and density, it will exert a pressure of 15 pounds to the inch. By subjecting water to a greater pressure, its boiling point is raised, and the elastic force of the steam is increased. The Marcet's globe illustrates this principle. It consists of a metallic globe (Fig. 192), which is furnished with a long glass tube and scale, T, a stopcock, S, and a thermometer, A, whose bulb is inside the globe. In the bottom of the globe is a little mercury, into which the end of the tube, T, dips, and above the mercury is a quantity of water. The water is boiled until the air is driven out of the open stop-cock. At this moment, the elastic force of the steam is just 15 pounds to the inch. The stop-cock is now closed: the thermometer at once shows a rise of temperature, and at the same time the mercury begins to rise in the tube, showing an increase in the force of the steam. When the temperature of the boiling water has reached 249.5°, the expansive force of the steam is equal to two atmospheres, or 30 pounds, to the inch, and at 306° it is five atmospheres, or 75 pounds to the inch.

If, now, this elastic force of steam can be made to act alternately upon opposite sides of a piston, it will push it back and forth, from one end of a cylinder to the other, with power enough to move any amount of other machinery. This is accomplished in the *steam-engine*.

The Boiler. — The boiler of a steam-engine is usually made of plates of wrought iron riveted together in the form of a cylinder. In the best forms, there are tubes which run lengthwise through the body of the boiler, through which the flame and hot gases from the fire may pass. The water in the boiler surrounds these tubes, and is rapidly heated by them. The steam thus formed in the boiler collects above the water, and by its pressure raises the boiling point, until, when its elastic force is sufficiently great, the steam is allowed to pass through a pipe to the cylinder.

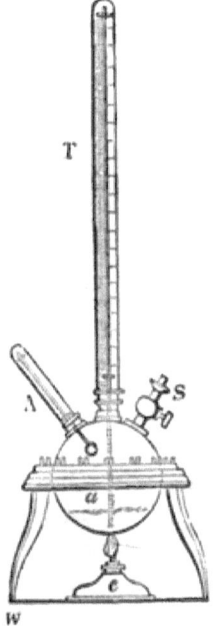

Fig. 192.

The Cylinder. — The arrangement of the cylinder and piston are shown in Fig. 193. The pipe which brings the steam from the boiler enters a box, *d*, from which two tubes lead, one to the top, the other to the bottom of the metallic cylinder, C, in which the piston, P, moves. Another tube leads from this box out into the air, or away to another vessel, where the steam, after having moved the piston, may be condensed. A sliding valve, *y*, is so arranged in the box as to always close one of the pipes leading to the cylinder, and leave the other open. If the upper tube is open, as represented in the figure, the steam will enter above the piston, and push it to the bottom of the cylinder; if the lower tube is open, the steam will enter below the piston, and push it to the top. In either case the steam on the opposite side of the piston will be pushed out of the cylinder, through the other tube and the pipe, O, leading from the cavity under the sliding valve. When the steam, entering through the lower tube, has pushed the piston to the top of the cylinder, the valve is pushed down to cover the end of that tube, leaving the end of the other uncovered, so that the steam may pass through it to act above the piston. By this means the piston will be alternately pushed back and forth from one end of the cylinder to the other.

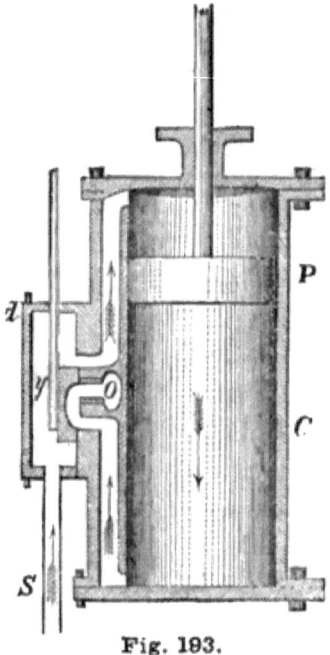

Fig. 193.

The Crank. — By this simple motion, back and forth, the piston turns a wheel by means of a crank. To the piston-rod, A (Fig. 193), another rod is joined by a hinge; the other end of this rod turns a wheel, from which motion may be communicated to others by bands or cogs.

Besides these three important parts of the steam-engine, there are numerous other appendages for particular purposes, such as a safety-valve attached to the boiler to regulate the pressure of steam in it; the governor, to regulate the supply of steam to the cylinder; the fly-wheel, a heavy wheel whose inertia causes the motion of the machinery to be steady.

Explain Fig. 194. — The picture (Fig. 194) shows how the piston-rod gives motion to machinery in another way.

In the first place, at the left, we see the cylinder with

NATURAL PHILOSOPHY. 299

one side cut away, so as to expose the piston, P, inside. The steam is supposed to be entering the valve-box at S, and to be going to the upper part of the cylinder, where it is pushing the piston down.

Next we observe that the piston-rod, A D, is fastened to one end of the large and strong lever, H K. As the piston goes down it pulls this end of the lever down, and throws

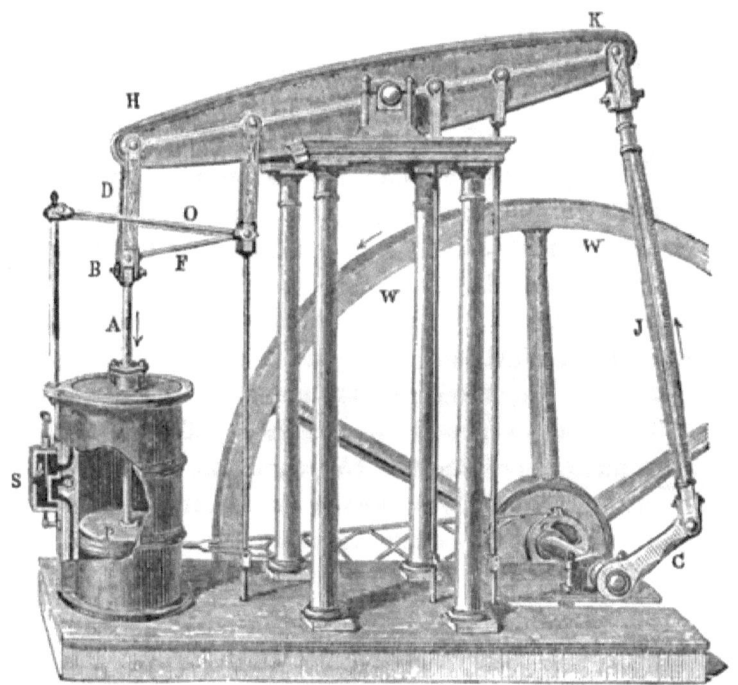

Fig. 194.

the other end, K, up. When the piston rises in the cylinder the piston-rod pushes the lever end, H, up, and throws the other end, K, down. Now, as the lever at K goes up and down, it pulls and pushes upon the strong arm, J, and in this way turns the crank, C. The large wheel, W W, fixed upon the axle, will thus be put in motion.

Where may we find Engines of this Form? — This form of engine is often used on steamboats. The great lever, H K, may be seen above decks moving alternately up and down when the steamer is in motion. It is called the "walking-beam." The strong arm, J, reaches down into the boat, and turns an enormous iron axle, which reaches quite through the boat from side to side, and has a *paddle-wheel* at each end.

High and Low Pressure Engines.—The different forms of steam-engines are almost as numerous as the machinists who make them, or as the variety of purposes to which they are applied. There are, however, two general classes, the *high-pressure* and the *low-pressure* engines.

In the high-pressure engines the steam, after moving the piston, is thrown out from the cylinder into the air. In the low-pressure engines, the steam, after moving the piston, is taken off to a vessel called the condenser, in which it is changed into water. The first is called *high* pressure, because the steam which moves the piston must push the steam from before the piston out into the air, which presses it back with a force of 15 pounds to the inch. To do this evidently requires a pressure of 15 pounds to the inch *higher* than in the other class, in which the steam escapes into a vacuum, and can, of course, exert no pressure against the piston.

SECTION IV.

REVIEW.

I.— SUMMARY OF PRINCIPLES.

Two forces, in opposite directions, will be in equilibrium when they do equal amounts of work. And work is measured by the product of the weight of the body moved by the vertical height through which it is lifted.

Machines are instruments with which forces may be applied to do work to better advantage than if they were to act directly upon the resistance itself.

The Lever, Wheel and Axle, Pulley, Inclined Plane, Wedge, and Screw are the simple machines out of which all compound machinery is made.

The power applied in machinery is sometimes moving water. In this case the power is applied by means of Water-Wheels.

The power applied is, very generally, the power of steam. In this case it is applied by means of a Steam-Engine.

II.—SUMMARY OF TOPICS.

109. The principle of work. — Machines. — Simple machines. — The law of equilibrium.

110. Levers. — Three classes. — Application of the principle of work. — Law of equilibrium. — The compound lever. — Applications of the lever.

111. The wheel and axle. — Acts on the principle of the lever. — Application of the principle of work. — The compound wheel and axle. — Wheels act by means of cogs. — By friction. — By bands. — Applications of the wheel and axle.

112. The pulley. — Fixed and movable. — Application of the principle of work to the fixed pulley. — To the movable pulley with one rope. — With separate ropes. — Applications of the pulley.

113. The inclined plane. — Power parallel to the length of the plane. — Power parallel to the base of the plane. — Applications of the inclined plane.

114. The wedge.

115. The screw. — Application of principle of work to the screw. — Applications of the screw. — Micrometers.

116. The undershot water-wheel. — The overshot wheel. — The breast wheel. — The turbine wheel.

117. The elastic force of steam. — The boiler of a steam-engine. — The cylinder. — The crank. — High and low pressure engines.

III. — PROBLEMS.

1. If a power of 10 pounds act upon the long arm of a lever, a distance from the fulcrum of 6 feet, what weight would it balance at a distance of 2 feet on the other side of the fulcrum?

Ans. 30 pounds.

2. In a lever of the second class, the power, 3 pounds, is at a distance of one foot from the fulcrum: what weight will it balance at a distance of one inch from the fulcrum?

Ans. 36 pounds.

3. In a compound lever, the long arms are 4 feet, 5 feet, and 6 feet in length; the short arms are 1 foot, 2 feet, and 3 feet long: a weight of 2,000 pounds is to be balanced: how much power must act upon the first long arm?

Ans. 100 pounds.

4. A power of 10 pounds lifts a weight of 500 pounds by means of a lever whose short arm is one foot long: how long is the long arm of the lever? *Ans.* 50 feet.

5. If the 500 pounds in the last example is to be lifted 2 feet, how far must the power move to do it?

Ans. 100 feet.

6. The radius of a wheel is 30 inches; of its axle, 5 inches; a power of 100 ounces is exerted upon the wheel: how much weight will it balance at the axle?

Ans. 600 ounces.

7. Three wheels and axles are combined, as shown in Fig. 178; the radius of each wheel is 20 inches; of each axle, is 4 inches; a power of 2 pounds acts on the first wheel: what weight will it balance on the last axle?

Ans. 250 pounds.

8. A force of 16 pounds is applied to the last axle (Fig. 178), and moves at the rate of 10 inches a second: how much weight at the first wheel would balance it at rest? and how much slower will it go when in motion?

Ans. .128 pound; $\frac{1}{125}$ as fast.

9. With a single movable pulley a stone weighing 350 pounds is to be lifted: what power must be exerted?

Ans. 175+ pounds.

10. With the single movable pulley, shown in Fig. 195, what power at P would balance a weight of 250 pounds at W? Ans. 83⅓ pounds.

11. If the weight, W (Fig. 195), is lifted by the power, how far would the power move to lift the weight one foot? Ans. 3 feet.

12. In a system of 4 movable pulleys, with a single rope, what power would be needed to balance a weight of 500 pounds?

Ans. 62½ pounds.

13. Suppose each of the 4 pulleys has a separate rope, what power would then be needed?

Ans. 31¼ pounds.

Fig. 195.

14. An inclined plane, 6 feet in length and 2 feet high, is used to put a barrel of flour upon a cart. The barrel weighs 196 pounds: how much force must a man exert, pushing parallel to the length of the plane?

Ans. 65⅓+ pounds.

15. If the base of the plane were 5 feet, its height 2 feet, and the man pushes parallel to the base, how much force must he exert to lift the barrel of flour?

Ans. 78⅖+ pounds.

16. The distance between the threads of a screw is one inch, and the power of 25 pounds moves in a circle of 3 feet in circumference: how much weight will it balance?

Ans. 900 pounds.

17. A power of 20 pounds, by means of a screw, exerts a pressure of 800 pounds. The threads are one-half inch apart: what is the circumference of the circle in which the power moves? Ans. 20 inches.

INDEX.

(The numbers refer to pages.)

Absolute temperature	61
Absolute weight	48
Acceleration	76
Action of heat	135
Adhesion defined	13
between solids	13
between liquids and solids	14
illustrated by experiment	13
Air, heated by convection	134
undulations in	106
vibrations of	102
weighing of	36, 46
Air-pump	45
Air thermometer	139
Alcohol thermometer	129
Apparatus for induction	261
Archimedes, principle of	34
Armature	241
Artesian wells	30
Ascent of a balloon	48
Astatic needle	260
Atmospheric pressure	48
shown by the barometer	52
depends on water-vapor	54
Atmosphere, pressure of one	51
density of	59
Atom defined	7
Attraction	8, 16, 17
Atwood's machine	71
Avogadro's law	61
Balance	279
Balloon, ascent of	48
Barometer	52

INDEX.

Barometer, corrections 53
        Forlin's 53
        predicts changes in the weather 54
Bell, vibrations of 101
Bell telephone 265
Black lines of the spectrum 198
Boiling point 142
        depends on pressure 143
        depends on purity of the liquid 142
        depends on the nature of the vessel 143
        expansion at, increases 144
        temperature at, constant 144
Boyle's Law 57
Breast-wheel 294
Bright lines in the spectrum 198
Camera-obscura 214
Capillarity 14
        familiar examples of 15
        illustrated by experiment 14
        law of 15
Center of gravity 81
Centigrade thermometer 138
Central forces 84
Charles' Law 60
Chemical action 259
Chemism 15
Cohesion 12
Color of bodies 204
   depends on rapidity of vibrations 207
   of clouds 205
   of the sky 204
   produced by interference of light 206
Components 78
        in a given direction 79
        in the same direction 80
Composition of forces 77
Compressibility of gases 44
        of liquids 24
Concave lenses 193
Concave mirrors 175
Conduction of heat 132
Conductivity 133
Conductors 133
Convection 134
Convex lenses 188
Convex mirrors 177
Cords, vibrations of 98

Cords, vibrations, laws of	97
rate of, determined	99
rate invariable	101
undulations of	103
Crystalline form	23
artificial	24
in nature	23
Curved motion	83
Decomposition of light	195
Diatonic scale	155
Dispersion of light	194
Dispersive power	196
Diffusion of heat	200
Diffraction of light	208
Divisibility	6
Double refraction	216
Ductility	4, 23
Dynamic theory of heat	126
Echo	152
Elasticity	3
of gases	45
of liquids	24
Electrical machine, frictional	225
the Holtz	233
Electric units	270
Electricity	224
applied to registering vibrations	99
conductors of	229
current	247
detected by electroscopes	227
dynamic	247
effects of	240, 254
electro-motive force	230
evolved by friction	224
evolved by chemical action	248
evolved by induction	233, 261
frictional machine for producing	225
Holtz machine for producing	233
induction	229, 261, 263
insulation	230
intensity of	252, 254
Ohm's law	252
polarization	230, 243
potential	229
quantity of	254
resistance to	252
strength of current	254

Electricity, units of 270
Electro-chemical telephone 265
Electrolytes 259
Electro-motive force 230, 252
Electroscopes 227
Energy 114
    chemical 122
    conservation of 123
    defined 118
    electrical 122
    kinetic 120
    measure of 120
    mechanical 121
    molecular 121, 132
    of the sunbeam 200
    potential 121
    radiant 122, 169
    recognition of, by the senses 126
    relation of, to mass and velocity 118
    relation of, to work 119
    transmission of 123
    transmutation of 123
    undulatory 148
    varieties of 121
Ether 128
Evaporation 142
Expansibility of gases 44
Expansion of gases by heat 137
    of liquids 136
    of solids 136
    temperature measured by 137
Extension 1
Eye, description of the 214
Fahrenheit's thermometer 138
Falling bodies 71
    analysis of the motion of 74
    formulas for 75
    laws of 75
Farad 271
Foci 176
Foot-pound 116
Force 115
    attractive or repellant 8
    centripetal 84
    centrifugal 84
    constant 71
    impulsive 70

INDEX. 309

Force in solids, liquids, and gases	20
producing motion	67
Forces, central	84
composition of	78
molecular, action of	20
of nature	10
resolution of	79
Fortin's barometer	53
Fulcrum	276
Fundamental ideas	16
explain the three forms of matter	20
explain the phenomena of motion	67
Galvanometer, astatic	260
Gases, characteristic properties of	43
expansion of	44, 137
kinetic theory of	51
molecular forces in	20
specific gravity of	36
the three laws of	57
volume of, depends on pressure	58
volume of, depends on temperature	60
Glass, ductility of	4
Gold, malleability of	4
Gravitation	10
laws of	10
not limited to the earth	12
Gravity, center of	81
specific	35
Hardness	21
Heat	132
a manifestation of energy	126
conduction of	132
convection	134
diffusion of	200
dynamic theory of	126
effects of	135
evolved by blows	121
evolved by chemical energy	122
evolved by electricity	254
latent	139, 140
mechanical equivalent of	124
restoration of	145
sensible	139, 140
specific	140
Ice, contraction of, by heat	141
Iceland spar	216
Impenetrability	2

Images 178
    by concave lenses 193
    by concave mirrors 180
    by convex lenses 190
    by convex mirrors 183
    by plane mirrors 179
Inclined plane 288
        applications of 290
Indestructibility 2
Index of refraction 186
Induction 229, 230
    by a current of electricity 261
    by a magnet 263
    by frictional electricity 229
    coils 262
    Faraday's theory of 232
Inertia 7
Interference 104
    of air-waves 108
    of light 205
    of sound 205
    of water-waves 105
Intervals in music 155
Kinetic energy 120
Kinetic theory of gases 51
Latent heat of water 142
Law of Avogadro 61
  of Boyle 57
  of capillarity 15
  of Charles 60
  of conservation of energy 123
  of electrical attraction and repulsion 228
  of electrical induction 231
  of gravitation 10
  of intensity of light 171
  of equilibrium 275
        for the inclined plane 289, 290
        for the lever 278
        for the pulley 286, 287
        for the screw 292
        for the wedge 291
        for the wheel and axle 281
  of magnetic attraction and repulsion 242
  of Marriotte 57
  of Ohm 252
  of reflection of light 172
  of reflection of sound 152

INDEX. 311

Law of refraction of light 186
   of transmission of light 169
   of velocity of sound 148
Laws of falling bodies 75
   of motion 67
Lenses 187
    effect of concave 189
    effect of convex 188
    images by 190
Levers 275
    applications of 279
    classes of 276
    compound 278
    law of equilibrium 278
    principle of work applied to 276
Light 169
    a manifestation of energy 129
    an undulatory motion 128
    analogous to sound 205
    definition of 129
    diffraction of 208
    dispersion of 194
    double refraction of 216
    intensity of 171
    interference of 205
    polarization of 217
    rays of 169
    reflection of 172
    refraction of 184
    transmission of 170
    velocity of 170
    wave-lengths of 206
Line of direction 82
Liquefaction 141
Liquids, characteristic property of 24
    convection in 135
    compressibility of 24
    elasticity of 24
    expansion of, by heat 136
    mobility of 25
    molecular forces in 20, 25
    pressure of 26
    specific gravity of 36
Loadstone 241
Luminous bodies 128
Machinery 274
Machines 275

INDEX.

Magic-lantern 213
Magnetic needle 245
     astatic 260
     affected by a current 260
     dip of 247
     variations 246
Magnetism 241
Magneto-electricity 264
Magneto-electric machine 264
Magneto-telephones 265
Magnets 241
Malleability 4
   of metals 22
Marriotte's law 57
Mass 6
  distinguished from weight 12, 114
  measure of 113
Measure of altitudes 54
   of electric current 270
   of energy 118
   of extension 2
   of force 115
   of mass 114
   of wave-length 206
   of work 116
Mechanical powers 275
Melting point 141
Metals, ductility of 23
Metric measures . . . . x, 2
Micrometers 293
Microphone 268
Microscopes 209
Minute division 6
Mirrors 174
Mobility in liquids 25
   in gases 47
Molecular forces 20
Molecule 6
Momentum 129
Motion 67
  application of the fundamental ideas 67
  curved 83
  elements of 69
  Newton's laws of 67
  of air 90
  of a falling body 71
  of liquids 86

Motion produced by a single force	67
produced by two or more forces	77
uniform	69
uniformly accelerated	71
Musical flames	163
Musical instruments	161
Musical sounds	153
intensity of	157
pitch of	155
quality of	158
Nature, changes of condition in	20
crystalline forms in	23
forces of	10
Natural philosophy defined	5, 17
Newton's rings	207
Nodes and segments	105
Ohm, the	270
Ohm's law	252
Opera-glass	211
Optical instruments	209
Organ-pipes	162
Oscillation, center of	96
Overshot wheel	294
Overtones	159
Parallelogram of forces	77
Pendulum, described	93
center of oscillation	96
laws of the	93
used to measure time	97
used to determine the form of the earth	97
vibrations of	94
Phonograph	159
Photometry	171
Physics, defined	5, 130
Pitch of sound	155
Polariscopes	218
Polarization, electrical	229
of the battery	250
of light	216
Potential, electrical	229, 269
Potential energy	121
Point of application	80
Press, hydrostatic	42
Pressure, a unit of	51
equal transmission of, by water	41
of the atmosphere	48
of liquids	26

Principle of stability	82
Prisms	194
Projectiles	85
Properties, chemical	4
of matter	1
physical	4
Pulley, applications of	287
classification	284
principle of work applied to	285
Pump, air	45
forcing	56
suction	55
Sprengel, principle of the	90
Radiometer	51
Rainbow	202
Range or random	85
Réaumur's thermometer	138
Reflecting telescopes	212
Reflection from rough surfaces	183
of sound	151
of light	172
Refraction	184
by lenses	187
by prisms	194
double	216
index of	186
laws of	185
Refracting telescopes	210
Register, the electric	99
Registering vibrations	99
Repulsion	8
molecular	9
Resolution of force	79
Resultant of forces	78
Ritchie's induction coil	262
Ruhmkorff coil	262
Screw	291
applications of	293
micrometer	293
Sensitive flames	165
Siphon	56
Siren	154
Solids, adhesion between	13
characteristic properties of	21
expansion of by heat	136
loss of weight of, in water	33
no convection in	135

Solids, specific gravity of	38
Sound	127
compound	158
distance measured by	151
effect of temperature on	151
formula for velocity of	151
intensity of	157
laws of velocity of	150
musical	153
pitch of	155
quality of	158
reflection of	151
transmission of	149
velocity of	149
waves	148
Specific gravity	35
of gases	36
of liquids	36
of solids	38
table of	40
Spectroscope	197
Spectrum analysis	199
a pure	197
invisible parts of a	199
the solar	196
Springs	29
Steam, heating by	145
tension of	296
Steam-engine	297
Steelyard	279
Stereopticon	213
Stringed instruments	161
Telephone, Bell's	265
Edison's	267
Telescope	210
astronomical	211
Galileo's	211
Herschellian	212
terrestrial	212
Temperature	137
absolute	61
measured by expansion	137
Tempering	22
Tenacity	22
Thermo-electricity	268
Thermo-electric pile	269
Thermometers	138

INDEX.

Thermometers, relation of scales	146
varieties of	138
Time measured by the pendulum	97
Tone, continuous	153
Trade-winds	91
Transmission of energy	123
of heat	201
of light	170
of sound	149
Transmutation of energy	123
Turbine wheel	294
Undershot water-wheel	293
Undulations	103
defined	104
in air	106
in cords	103
in water	105
period of	104
phase of	104
Undulatory energy	148
Uniformly accelerated motion	71
Uniform motion	69
Unit of pressure	51
of work	116
Units in electrical measures	270
Vacuum, by air-pump	45
by stream of water	90
Vaporization	142
Velocity defined	69
of a jet of water	86
uniform	69
uniformly accelerated	71
Vibration	92
amplitude of	94
of air	102
of a bell	101
of cords or wires	97
of a pendulum	93
of water	101
phase of	103
producing hearing	127
rate of invariable	101
registered by electricity	99
transmission of	103
transverse and longitudinal	108
Vision	128
Voice, to record the	160

INDEX. 317

Voice, to reproduce the	160
Volt	271
Water, composition of	7
decomposition of	259
discharged from an orifice	87
latent heat of	142
rising in pipes	28
surface of	27
supply to cities	29
vibrations of	101
Water-power	293
Wave, defined	104
front	169
length	104, 206
period	104
phase	104
Waves of air	106
of water	105
interference of	104
Wedge	291
Weighing air	46
Weight, defined	11
absolute	48
distinguished from mass	12, 115
laws of	11
not confined to bodies on the earth	12
of air	45
the resultant of parallel forces	81
Welding	13
Wells, artesian	30
Wheelbarrow	280
Wheel and axle	280
acts as a lever	281
applications of the	283
compound	282
law of equilibrium	281
turned by bands	283
by cogs	282
by friction	282
Wind	91
Wind instruments	162
Work, defined	116
measure of	116
principle of	274

www.ingramcontent.com/pod-product-compliance
Lightning Source LLC
Chambersburg PA
CBHW030738230426
43667CB00007B/761